Hungarian Mathematical Olympiad (1964–1997)

Problems and Solutions

Mathematical Olympiad Series

ISSN: 1793-8570

Series Editors: Lee Peng Yee *(Nanyang Technological University, Singapore)*
Xiong Bin *(East China Normal University, China)*

Published

Vol. 20 *Hungarian Mathematical Olympiad (1964–1997):*
Problems and Solutions
by Fusheng Leng (Academia Sinica, China),
Xin Li (Babeltime Inc., USA) &
Huawei Zhu (Shenzhen Middle School, China)

Vol. 19 *Mathematical Olympiad in China (2019–2020):*
Problems and Solutions
edited by Bin Xiong (East China Normal University, China)

Vol. 18 *Mathematical Olympiad in China (2017–2018):*
Problems and Solutions
edited by Bin Xiong (East China Normal University, China)

Vol. 17 *Mathematical Olympiad in China (2015–2016):*
Problems and Solutions
edited by Bin Xiong (East China Normal University, China)

Vol. 16 *Sequences and Mathematical Induction:*
In Mathematical Olympiad and Competitions
Second Edition
by Zhigang Feng (Shanghai Senior High School, China)
translated by: Feng Ma & Youren Wang

Vol. 15 *Mathematical Olympiad in China (2011–2014):*
Problems and Solutions
edited by Bin Xiong (East China Normal University, China) &
Peng Yee Lee (Nanyang Technological University, Singapore)

Vol. 14 *Probability and Expectation*
by Zun Shan (Nanjing Normal University, China)
translated by: Shanping Wang (East China Normal University, China)

The complete list of the published volumes in the series can be found at
http://www.worldscientific.com/series/mos

Vol. 20 | Mathematical
Olympiad
Series

Hungarian Mathematical Olympiad (1964–1997)

Problems and Solutions

Fusheng Leng
Academia Sinica, China

Xin Li
Babeltime Inc., USA

Huawei Zhu
Shenzhen Middle School, China

W⊖ World Scientific

EW JERSEY · LONDON · SINGAPORE · BEIJING · SHANGHAI · HONG KONG · TAIPEI · CHENNAI · TOKYO

Published by

World Scientific Publishing Co. Pte. Ltd.

5 Toh Tuck Link, Singapore 596224

USA office: 27 Warren Street, Suite 401-402, Hackensack, NJ 07601

UK office: 57 Shelton Street, Covent Garden, London WC2H 9HE

Library of Congress Cataloging-in-Publication Data
Names: Leng, Fusheng, author. | Li, Xin, author. | Zhu, Huawei, author.
Title: Hungarian Mathematical Olympiad (1964-1997) : problems and solutions /
    Fusheng Leng, Academia Sinica, China, Xin Li, Babeltime Inc., USA,
    Huawei Zhu, Shenzhen Middle School, China.
Description: New Jersey : World Scientific, [2023] |
    Series: Mathematical Oympiad series, 1793-8570 ; volume 20
Identifiers: LCCN 2022037350 | ISBN 9789811255557 (hardcover) |
    ISBN 9789811256363 (paperback) | ISBN 9789811255564 (ebook)
Subjects: LCSH: International Mathematical Olympiad. | Mathematics--Problems, exercises, etc. |
    Mathematics--Competitions--Hungary.
Classification: LCC QA43 .L565 2023 | DDC 510.76--dc23/eng/20220809
LC record available at https://lccn.loc.gov/2022037350

British Library Cataloguing-in-Publication Data
A catalogue record for this book is available from the British Library.

For any available supplementary material, please visit
https://www.worldscienti ic.com/worldscibooks/10.1142/12815#t=suppl

Printed in Singapore

To our mentor

Prof. Zonghu Qiu

and

To the memory of

János Surányi

the Hungarian author
on whose work this book is based

In Memory of
JÁNOS SURÁNYI

My father was born on May 29, 1918 and passed away on December 8, 2006, in Budapest, Hungary. He grew up in a cultured family. Margit Gráber (1895–1993), an aunt of his, was a well-known painter. Ender Bíró (1919–1988), a cousin of his, was an important biochemist. He was the founding Head of the Department of Biochemistry of the Eötvös Loránd University in Budapest, and the first person to translate parts of James Joyce's *Finnegans Wake* into Hungarian.

As a youngster, my father attended the Circle organized by the thinker Sándor Karácsony (1891–1952). Another mentor of my father's in those early days was the thinker and philosopher Lajos Szabó (1902–1967), with whom he had correspondences exploring Gödel's Incompleteness Theorem, to the mutual benefit of both parties.

My father entered the University of Szeged in 1936. Here he was exposed to mathematical logic, linking his favorite areas: mathematics and philosophy. He became a student of László Kalmár (1905–1976), the leading Hungarian authority on the subject. In 1943, he obtained his doctorate from the University of Szeged.

During the Second World War, the Hungarian government sided with the Nazis. My father was taken into the Hungarian army as an unarmed labourer. Fortunately, his boss was secretly against the war, and often listened with my father's group to Radio London, which was of course strictly forbidden. Being Jewish, he was later turned over to the Germans, and this time, his boss was a trigger-happy teenager. Somehow, my father managed to survive.

After the war, my father returned to the University of Szeged to work as an assistant of Kalmár. In 1951, he became a faculty member at the Eötvös Loránd University. In 1953, he obtained a degree called the "Candidature". In 1957, he became a Doctor of the Academy of Science.

The main results of his dissertations were on the "Decision Problem", which represented the completion of this area of research. They were summarised in his book *Reduktionstheorie des Entscheidungsproblems in Prädikationskalkül der ersten Stufe*, published in Budapest and Berlin in 1959. His other research areas were number theory and graph theory.

In 1960, my father was promoted to a Professorial Chair. From 1976 till his retirement in 1988, he was the Head of the Department. Besides books and publications, he presented his results at national and international congresses and conferences.

While his research work was solid, my father was best remembered as the **driving force** behind the revolution in mathematics teaching in Hungary.

This became his main interest in the latter part of his career. He had publications in this area in many languages. He was twice the president of the International Mathematical Olympiad when the event was held in Hungary.

I remember that when I was very young, I often saw talented secondary school students coming to my father to discuss mathematical issues. His friend Paul Erdős, when staying in Hungary, was willing to travel a hundred miles to meet these talented students.

The two wrote a book *Topics in the Theory of Numbers* (first published in 1960; revised editions followed almost till my father's death; an English translation appeared in 2003). The academician Miklós Laczkovich once told me, "I read my copy of this book so much that it fell apart."

At the memorial service to my father, László Lovász, a winner of the Wolf Prize and the Abel Prize, gave a speech. The following remark of his reflected the thoughts of many: "I only really understood what mathematics is from the beautiful proofs of this book."

Then he added, "As a student, I often visited János in his apartment, where he taught me number theory, presented me some problems and gave me things to read. In his explanations, he always concentrated on the essence of the argument; he showed how proofs emerge, and how one could have discovered them."

My father took up the actual writing of the bulk of the text. Even when Erdős was visiting other continents, the two of them continued to work together by correspondence in drawing up the detailed list of topics and results to be discussed. I still have a few of these letters.

Some of the theorems in the book were too difficult for secondary school students who still lacked a deeper understanding of mathematics. My father used the appropriate language to smooth over their learning process.

The modern state of Hungary, a small land-locked country, was born after the First World War. After the Second World War, it fell under the influence of the former Soviet Union, until just before the latter's dissolution.

Under authoritarian governments, the period from 1919 to 1990 were difficult times for Hungarian academics. Disciplines such as social and biological sciences were heavily politicized. As a result, many talented people drifted towards mathematical and physical sciences, in search for a freer environment to conduct their research and teaching.

The years between 1945 and 1948 were perhaps the period in Hungary, during these seven decades, when people had a little more freedom. My father took advantage of this opportunity and in 1947, with the help of Paula Soós, he restarted the legendary *KöMaL* (Secondary School Mathematics and Physics Journal). This publication had already played an important role in the development of mathematics in Hungary in the early twentieth century. My father's generation grew up on the problems and articles in it. He remained its main editor until 1970.

An amusing incidence was that the publication was advertised as a medium for students to *socialize* in mathematics. This pleased the governing *Socialist* Party and allowed the journal to carry on its glorious tradition. A very large number of students from all over the country sent in solutions every month to the problems posed in this journal. Some post offices had students working there till midnight on the deadline-days.

In teaching mathematics to secondary school students and even younger children, my father had been inspired by his mentor Kalmár. It was Hungary's good fortune that world-class mathematicians, such as Erdős, Alfred Rényi and Tibor Gallai, were similarly committed. I still have letters of my father to Szabó in which he related his discussions with Rényi about the importance of school education.

When Rényi was the Director of the Mathematical Institute of the Hungarian Academy of Science, he provided space for my father to organize lectures and discussions on new methods in mathematics teaching. Later Rényi also managed to found a formal Study Group in the Institute led by my father.

The most important thing for these mathematicians was *to turn mathematics into a positive experience for students*. This clearly required the modernization of the teaching of the subject.

For my father, the experience of mathematics consisted of *a combination of logical rigour and a playful approach* — in his eyes the main purpose of mathematics teaching was to convey this unity, and at the same time to develop independent thinking in students.

Of course, the tone of the lessons must be in the spirit of freedom. It was necessary to move away from the traditional treatment of the subject, largely based on the old Prussian way, consisting mainly of routine exercises. From the sixties, despite strong opposition by the government, a movement towards a free-thinking and discovery-based approach — one that allows for the emergence of mathematical notions — spread to the whole of mathematics teaching.

It became necessary to have contact with the leading figures of the new mathematics movement in Western countries, and gradually, connections were established. My father was particularly close to Emma Castelnuovo, Anna Zofia Krygowska, Willy Servais and Klaus Härtig, with whom he also shared common interest in logic and number theory.

My father began to take part and also give lectures on Colloquiums of the International Commission on Mathematical Instruction (ICMI), and this greatly strengthened the movement at home. Some articles and new "experimental" textbooks were published. In 1971–1974, he became the first vice-president of the ICME, and in 1975–1978 a member-at-large. Their website paid tribute to my father: "Around the time of Surányi's vice-presidency, there was an explosion of the international events that shaped the future developments of ICMI. He was an active participant."

We now turn from mathematics education for the general students to that for the special students, such as those who had visited my father to discuss mathematical issues.

In one secondary school in 1962, a so-called "special class in mathematics" was permitted with twice as much mathematics lessons as in other classes. Later on, more such classes began to appear in the country. But without a curriculum or textbooks, this had not been easy. My father was the leader in this movement, with the help of a lot of young university lecturers. It was hard and very intensive work. Many people went out to those special classes to discuss with the teachers lesson plans and how to continue the work. Almost all the Hungarian team members for the International Mathematical Olympiad were chosen from these classes.

I attended one of these classes. We had an inspirational teacher Köváry whose name meant "Stone Castle". My father liked him very much. His former students gave him the nickname of Kavics, which meant "Pebbles". Eventually everyone, including himself, called him that. From him we learnt things that had never been taught in the country before. I can remember him stating loudly and smiling: "*Holnap jön Apu*". It meant that the classroom would have unknown adults observing our excited and exciting learning process.

My father's only regret was that while such special classes represented his ideal way of teaching, too much energy and resources were taken up for the mathematics education of too few students in the country. Much more needs to be done.

As is well documented, the Hungarian secondary school mathematical competition named after Loránd Eötvös is the oldest in the world that is organized on a national scale. It started in 1894, and was interrupted in 1919–1921, 1944–1946 and 1956 by world political events. After the Second World War, my father, then 27 years old, restarted the competition and renamed it after József Kürschák. My father was the chair of its Organising Committee, and had a major influence on its direction and outlook.

My father always emphasized that both the problems and the solutions should be inventive, and should not require more than what was taught in an average secondary school. He also insisted that solutions must be thorough and easy to follow.

Details of the annual contest were summarized in the volumes of *Matematikai Versenytételek*. My father co-authored Volume I and Volume II with other mathematicians, and completed the next two volumes on his own. These books have been translated into several languages, including Russian and Japanese.

The book you are holding is based on Volume III and Volume IV, but it is not a direct translation. In particular, it presents only one solution to each problem, whereas in the original work, there are multiple solutions. In an attempt to appeal to a Western readership, the contents have been slightly revised as well as reorganized. There are also some additional material. I am happy with how it has turned out, and I believe so would my father.

László Surányi
Budapest
June 2022

Foreword

If you are reading this, you are probably interested in mathematics competitions, either as an eager student looking for training material, or an experienced tutor looking for teaching resources. Whichever the case, I strongly recommend that you purchase this book.

When the International Mathematical Olympiad started in Romania in 1959, I was still in elementary school in Hong Kong. Thus I had no opportunity to participate as a contestant. I had served as the deputy leader of the U.S.A. team from 1981 to 1984 (**Murray Klamkin** was the leader), the leader of the Canadian team in 2000 and 2003, the chair of the Problem Committee in 1995 when the I.M.O. was held in Canada, and a member of the Problem Committee in 1994, 1998 and 2016.

I first learnt of the I.M.O. from my mentor **William Brown**, when I was an undergraduate student at McGill University in Montreal. He had played an important role in establishing the Canadian Mathematical Olympiad. Apart from being an accomplished mathematician, he was also an amazing linguist, with great proficiency in Arabic and Hungarian, among other exotic languages. He had often told me about the glorious mathematics traditions of Hungary.

As it happened, my first trip to Europe was to the 1982 I.M.O. in Budapest, Hungary. I bought an English-Hungarian dictionary and a Hungarian-English dictionary. Unfortunately, I was unable to make as much progress as my mentor. Nevertheless, it piqued my interest in the general and mathematical history of the country.

Between 1867 and 1918, Hungary was part of the dual monarchy in the Austro-Hungarian Empire. After the First World War, it had shrunk to a small land-locked country.

With limited resources and manpower, Hungary continues to produce mathematical talents at an incredible rate. It is worthwhile finding out how this can happen.

In 1891, Baron Loránd Eötvös, a prominent physicist, initiated the founding of the Mathematical and Physical Society. In 1894, the Baron became the Minster of Religion and Public Education. The Eötvös Mathematics Competition was organized in his honor. Some months before the first competition, a *Mathematical Journal for Secondary Schools* made its first appearance. After Eötvös's death in 1919, the society was renamed the Eötvös Loránd Mathematical and Physical Society.

In 1947, the society had split into two, The Eötvös Loránd Physical Society and the Bolyai János Mathematical Society. János Bolyai was one of the founding fathers of Non-Euclidean Geometry. In 1949, the competition became the Kürschák Mathematics Competition, named after the mathematician József Kürschák.

All these phenomenal events started more than a century ago, way before anywhere else in the world. Thus a solid foundation had been laid. Since its inception, the Bolyai János Society continues as the guiding light in Hungarian mathematics. It has taken over the editing of the journal which had had a checkered history. From this point on, the journal, known by its acronym KÖMAL, is put on a firm footing.

Of the three two-time first-prize winners of the Kürschák Mathematics Competition during the period covered by this book, I had met **Gábor Tardos** at the I.M.O. in 1981, and corresponded with **Géza Kós**. Géza has become a prominent member of an international strike force that augments the Problem Committees of the I.M.O. in many countries. Regrettably, I have not crossed paths with **Tamás Fleiner**.

I have authored, edited and read many competition books. The book you are reading has several outstanding features. The problems are from a Hungarian treasure trove. The general discussions on the six topics can serve as a textbook in problem solving, with the actual contest problems as illustrative examples. The detailed solutions are carefully presented. The additional problems in the Appendix take you from competitions to various mathematical topics beyond. It is an amazing package.

Andy Liu
Edmonton
June 2022

Preface

There are quite a lot of problem books in the literature nowadays. Many such books have titles like "The Best Problems from Around the World" or "One Hundred of My Favorite Problems". The unfortunate thing is that about half of the book is already known to the reader. Moreover, this is often the better half of the book.

Our book is different in that it is based on a single source. We make no selection on our own apart from choosing the source in the first place, and what a fantastic source it is!

The Eötvös-Kürschák Mathematics Competition is the oldest high school mathematics competition in the world organized on a national scale, dating back to 1894. It was originally named after Baron Loránd Eötvös, who was a physicist. After the Second World War, the physics competition inherited this name and the mathematics competition was renamed after József Kürschák, a mathematician.

This competition is often considered to be the **Hungarian Mathematical Olympiad** by people outside the country. Volumes 1 and 2 in Hungarian, covering the H. M. O. from 1894 to 1963, have been translated and published in English as Hungarian Problem Books I, II, III and IV by the Mathematical Association of America. János Surányi was the sole author of Volumes 3 and 4, covering the H. M. O. from 1964 to 1997.

This book is *not* a direct translation of János' work. Its intended audience is an eager student training for mathematics competitions. The assumed background is basic knowledge in arithmetic, algebra and geometry up to the junior high school level, along with some rudimentary problem-solving techniques such as mathematical induction. This book is particularly suitable for self-study, though it can also serve as a resource for a teacher training their students for mathematics competitions.

The value of this book lies in the competition problems. They are all important, instructive and interesting. The statements are given in Part I in chronological order. Full solutions to the problems are given in Part III, also in chronological order. These were worked out by Fusheng Leng and Xin Li while they were still high school students. We present only one solution to each problem, and encourage the reader to find alternative approaches.

Part II contains the problems reorganized by topics into six sets. Within each set, the problems are presented in roughly ascending order of difficulty, according to the judgment of the authors. Of course, there will not be any easy problems in the competition, but some are definitely more accessible than others. A few problems require greater mathematical sophistication. A general discussion of the topic is followed by a specific discussion of each problem. These are prepared by Huawei Zhu.

Hungary and Russia are arguably the top two countries in problem proposing. Russia is a vast country and commands a large intellectual force. Hungary, on the other hand, is very small. Its strength comes from its glorious tradition to which we have referred. It would be impossible to talk about mathematics competition in Hungary without mentioning KÖMAL, which stands for *Középiskolai Matematikai és Fizikai Lapok*, or "Secondary School Mathematical and Physical Journal". It is the oldest publication of its kind in the world, also started in 1894. It is still going as strong as ever well into its second century.

The authors are grateful to all staff members of World Scientific, in particular **Lai Fun Kwong**, for their encouragement, advice and support.

Fusheng Leng, Research Assistant, Academia Sinica, Beijing;
Xin Li, Chief Technology Officer, Babeltime Inc., San Mateo;
Huawei Zhu, Principal, Shenzhen Middle School, Shenzhen,
June 2022

Contents

List of Winners

The Hungarian Mathematical Olympiad plays an essential role in identifying and nurturing mathematical talents in Hungary over the years. Among past winners, many have attained international fame.

László Lovász, who was on the winners' list twice, was a key figure in the Microsoft Corporation, but had since returned to Hungary. He had served as Director of the Mathematics Institute of Eötvös Loránd University during 2006–2011, and as the President of the Hungarian Academy of Sciences during 2011–2020.

József Pelikán, also on the 1965 winners' list, was a most important figure in the International Mathematical Olympiad movement. A prolific problem proposer and a terrific problem solver, he had served as the leader of the Hungarian team for many years and also as the Chair of the I.M.O. Advisory Board during 2002–2010.

The competition was originally intended for students who has just graduated from high school. In later years, some students still in the four-year high-school program were allowed to participate. In the list below, (2), (3) and (4) indicate respectively students entering the second, third and fourth years. Names in boldface indicate first prize winners. A dash indicates that no first prize was awarded that year. Note that Hungarian family names precede given names.

1964. Gerencsér László, Corrádi Gábor, Lovász László (3).

1965. Makai Endre, Lovász László (4), Pelikán József (4).

1966. Hoffmann György (4), Elekes György (4).

1967. —, Babai László (4), Mérő László (4), Szeredi Péter.

1968. Lempert László (3), Bajmóczy Ervin (2), Mihaleczky György (4).

1969. Fiala Tibor (2), Gönczi István (4), Bajmóczy Ervin (3), Csirmaz László, Lukács Péter (3), Nagy András (3), Pintz János.

1970. Bajmóczy Ervin (4), Ruzsa Imre (4).

1971. Komjáth Péter, Ruzsa Imre, Füredi Zoltán (4).

1972. Tuza Zsolt, Győri Ervin, Prőhle Péter (3).

1973. Kertész Gábor (4), Kollár János (4), Bacsó Gábor.

1974. —, Kollár János, Neumann Attila (4), Prőhle Péter, Spar -ing László (4), Kecskés Csaba, Markó Péter (4).

1975. Miklós Dezső, Jakab Tibor, Bodó Zalán (3), Lelkes András, Moussong Gábor (4).

1976. Magyar Zoltán (4).

1977. —, Knébel István.

1978. —, Cseri István (4), Csók Tibor, Sali Attila.

1979. Szegedy Márió, Hajnal Péter, Tardos Gábor (2), Bohus Géza (4), Szegedy Patrik (4), Varga Tamás, Beleznay Ferenc (4).

1980. Tardos Gábor (3), Bohus Géza, Szabó László, Horváth István (4).

1981. Tardos Gábor (4), Király Zoltán, Nacsa János (4).

1982. —, Magyar Ákos (3), Hetyei Gábor (4), Szabó Endre, Tardos Gábor, Csillag Péter (3).

1983. Birkás György (3), Erdős László (4), Hetyei Gábor, Kós Géza (2), Magyar Ákos (4), Pór István (4), Szabó Csaba, Törőcsik Jenő.

1984. Erdős László, Szabó Zoltán.

1985. Kós Géza (4), Csizmadia György (4), Montágh Balázs (4).

1986. Kós Géza, Benczúr András (4), Csom Gyula (3), Montágh Balázs.

1987. Drasny Gábor (4), Fleiner Tamás (3), Keleti Tamás (4), Tasnádi Tamás, Benczúr András, Dinnyés Enikő, Gács András, Lengyel Csaba, Lipták László, Rimányi Richárd.

1988. Fleiner Tamás (4), **Mándy Attila** (4), **Sustik Mátyás** (4); Bíró András (4).

1989. Fleiner Tamás; Sustik Mátyás; Balogh József (4).

1990. Pór Attila (3); Lakos Gyula (3); Harcos Gergely (4).

1991. Fleiner Balázs; Kálmán Tamás (3), Katz Sándor (3).

1992. Ujváry-Menyhárt Zoltán; Németh Ákos (3), Tichler Krisztián (4).

1993. Kálmán Tamás; Veres Gábor; Párniczky Benedek (4).

1994. Burcsi Péter (3), **Koblinger Egmont** (4); Futó Gábor, Szádeczky-Kardoss Szabolcs (4); Szobonya László (3).

1995. —; Braun Gábor (3), Burcsi Péter (4), Frenkel Péter (3), Gröller Ákos (4).

1996. —; Lippner Gábor (3), Braun Gábor (4), Kun Gábor (3).

1997. —; Gueth Krisztián (3), Lippner Gábor (4).

First Problem Index

Year	#1	#2	#3	Year	#1	#2	#3
1964	E8	A1	C6	1965	C3	F1	E10
1966	E4	C12	B2	1967	B12	F8	D5
1968	C2	F2	A5	1969	B3	D9	E1
1970	F16	A8	F12	1971	D7	F4	A9
1972	C4	A4	F15	1973	B13	E15	E5
1974	A3	F17	C11	1975	C1	D4	C13
1976	D10	A12	C17	1977	B8	D12	A2
1978	B10	F11	D17	1979	E7	C16	A15
1980	E2	B9	A17	1981	D3	A6	B15
1982	E16	B4	E11	1983	B11	C9	F3
1984	B14	E17	A16	1985	F7	B7	D8
1986	E9	C10	A7	1987	B1	E3	A13
1988	D2	C14	E14	1989	D16	B17	F6
1990	B5	D 15	A10	1991	C5	E6	F14
1992	C7	B16	F13	1993	B6	D11	C8
1994	D6	F9	F5	1995	E13	C15	D13
1996	D1	A11	F10	1997	E12	D14	A14

Second Problem Index

Set	A	B	C	D	E	F
#1	64-2	87-1	75-1	96-1	69-3	65-2
#2	77-3	66-3	68-1	88-1	80-1	68-2
#3	74-1	69-1	65-1	81-1	87-2	83-3
#4	72-2	82-2	72-1	75-2	66-1	71-2
#5	68-3	90-1	91-1	67-3	73-3	94-3
#6	81-2	93-1	64-3	94-1	64-1	89-3
#7	86-3	85-1	92-1	71-1	91-2	85-1
#8	70-2	77-1	93-3	85-3	79-1	67-2
#9	71-3	80-2	83-2	69-2	86-1	94-2
#10	90-3	78-1	86-2	76-1	65-3	96-3
#11	96-2	83-1	74-3	93-2	82-3	78-2
#12	76-2	67-1	66-2	77-2	97-1	70-3
#13	87-3	73-1	75-3	95-3	95-1	92-3
#14	97-3	84-1	88-2	97-2	88-3	91-3
#15	79-3	81-3	95-2	90-2	73-2	72-3
#16	84-3	92-2	79-2	89-1	82-1	70-1
#17	80-3	89-2	76-3	78-3	84-2	74-2

Part I: Problems
1964

Problem 1.
$PABC$ is a triangular pyramid with $AB = BC = CA$ and $PA = PB = PC$. Another triangular pyramid congruent to $PABC$ is glued to it along the common base ABC to obtain a hexahedron with five vertices such that the dihedral angle between any two adjacent faces is the same. Determine the ratio $PQ : BC$ where Q is the fifth vertex.

Problem 2.
At a party, every boy dances with at least one girl, but no girl dances with every boy. Prove that there exist two boys and two girls such that each of these two boy has danced with exactly one of these two girls.

Problem 3.
Prove that for any positive real numbers a, b, c and d,

$$\sqrt{\frac{a^2 + b^2 + c^2 + d^2}{4}} \geq \sqrt[3]{\frac{abc + bcd + cda + dab}{4}}.$$

1965

Problem 1.
Determine all integers a, b and c such that

$$a^2 + b^2 + c^2 + 3 < ab + 3b + 2c.$$

Problem 2.
Among eight points on or inside a circle, prove that there exist two whose distance is less than the radius of the circle.

1

Problem 3.
The base $ABCD$ and the top $EFGH$ of a hexahedron are both squares. The lateral edges AE, BF, CG and DH have equal length. The circumradius of $EFGH$ is less than the circumradius of $ABCD$, which is in turn less than the circumradius of $ABFE$. Prove that the shortest path on the surface of this hexahedron going from A to G passes through only the lateral faces.

1966

Problem 1.
Do there exist five points A, B, C, D and E in space such that

$$AB = BC = CD = DE = EA \quad \text{and}$$

$$\angle ABC = \angle BCD = \angle CDE = \angle DEA = \angle EAB = 90°?$$

Problem 2.
Let n be any positive integer. Prove that the first n digits after the decimal point of the decimal expansion of the real number $(5 + \sqrt{26})^n$ are identical.

Problem 3.
Do there exist two infinite sets of non-negative integers such that every non-negative integer is expressible as the sum of one element from each set in a unique way?

1967

Problem 1.
In a set of integers which contains both positive and negative elements, the sum of any two elements, not necessarily distinct, is also in the set. Prove that the difference between any two elements is also in the set.

Problem 2.
A convex polygon is divided into triangles by non-intersecting diagonals. If each vertex of the polygon is a vertex of an odd number of these triangles, prove that the number of vertices of the polygon is divisible by 3.

Problem 3.
The sum of the distances from each vertex of a convex quadrilateral to the two sides not containing it is constant. Prove that the quadrilateral is a parallelogram.

1968

Problem 1.
Prove that in any infinite sequence of positive integers, it is not possible for every block of three consecutive terms a, b and c to satisfy $b = \dfrac{2ac}{a+c}$.

Problem 2.
Let n be a positive integer. Inside a circle with radius n are $4n$ segments of length 1. Prove that given any straight line, there exists a chord of the circle, either parallel or perpendicular to the given line, that intersects at least two of the segments.

Problem 3.
Let $n > k > 0$ be integers. For each arrangement of n white balls and n black balls in a row, count the number of pairs of adjacent balls of different colors. Prove that the number of arrangements for which the count is $n - k$ is equal to the number of arrangements for which the count is $n + k$.

1969

Problem 1.
Let n be a positive integer. Prove that if $2 + 2\sqrt{28n^2 + 1}$ is an integer, then it is the square of an integer.

Problem 2.
Let the lengths of the sides of a triangle be a, b and c, and the measures of the opposite angles be α, β and γ respectively. Prove that the triangle is equilateral if

$$a(1 - 2\cos\alpha) + b(1 - 2\cos\beta) + c(1 - 2\cos\gamma) = 0.$$

Problem 3.
A $1 \times 8 \times 8$ block consists of 64 unit cubes such that exactly one face of each cube is painted black. The initial arrangement is arbitrary. In a move, we may rotate a row or a column of 8 cubes about their common axis. Prove that after a finite number of such moves, we can obtain an arrangement in which all the black faces are on top.

1970

Problem 1.
Let n be a positive integer. An n-gon on the plane is not necessarily convex, but non-adjacent sides do not intersect. Determine the maximum number of acute angles of this n-gon.

Problem 2.
Determine the probability that five numbers chosen at random from the first 90 positive integers contain two consecutive numbers.

Problem 3.
On the plane are a number of points no three of which lie on the same straight line. Every two of these points are joined by a segment. Some segments are painted red, some others are painted blue, while the remaining ones are unpainted. Every two of the points is connected by a unique polygonal path consisting only of painted segments. Prove that each unpainted segment can be painted red or blue so that in any triangle determined by these points, the number of red sides is odd.

1971

Problem 1.
A straight line intersects the sides BC, CA and AB, or their extensions, of triangle ABC at D, E and F respectively. Q and R are the images of E and F under 180° rotations about the midpoints of CA and AB respectively, and P is the point of intersection of BC and QR. Prove that

$$\frac{\sin EDC}{\sin RPB} = \frac{QR}{EF}.$$

Problem 2.
In the plane are 22 points where no three of which lie on the same straight line. Prove that they can be joined in pairs by 11 segments such that these segments intersect one another at least 5 times.

Problem 3.
Each of 30 boxes can be opened by a unique key. These keys are then locked at random inside the boxes, with one key in each. Two of the boxes are then broken open simultaneously, and the keys inside may be used to try to open other boxes. Keys retrieved from boxes thus opened may also be used. Determine the probability that all boxes may be opened.

1972

Problem 1.
Prove that $a(b-c)^2 + b(c-a)^2 + c(a-b)^2 + 4abc > a^3 + b^3 + c^3$, where a, b and c are the lengths of the sides of a triangle.

Problem 2.
In a class with at least 4 students, the number of boys is equal to the number of girls. Consider all arrangements of these students in a row. Let a be the number of arrangements for which it is impossible to divide it into two parts so that the number of boys is equal to the number of girls in each part. Let b be the number of arrangement for which there is a unique partition with this property. Prove that $b = 2a$.

Problem 3.
There are four houses in a square plot which is 10 kilometers by 10 kilometers. Roads parallel to the sides of the plot are built inside the plot so that from each house, it is possible to travel by roads to both the north edge and the south edge of the plot. Prove that the minimum total length of the roads is 25 kilometers.

1973

Problem 1.
Determine all integers n and k, $n > k > 0$, such that the binomial coefficients $\binom{n}{k-1}$, $\binom{n}{k}$ and $\binom{n}{k+1}$ form an arithmetic progression.

Problem 2.
For any point on the plane of a circle other than its center, the line through the point and the center intersects the circle at two points. The distance from the point to the nearer intersection point is defined as the distance from the point to the circle. Prove that for any positive number ϵ, there exists a lattice point whose distance from a given circle with center $(0,0)$ and radius r is less than ϵ, provided that r is sufficiently large.

Problem 3.
Let n be an integer greater than 4. Every three of n planes have a common point, but no four of these planes have a common point. Prove that among the regions into which space is divided by these planes, the number of tetrahedra is not less than $\frac{2n-3}{4}$.

1974

Problem 1.
When someone enters a library, she writes down on a blackboard the number of people already in the library at the time. When someone leaves a library, she writes down on a whiteboard the number of people still in the library. Prove that at the end of the day, the numbers on the blackboard are the same as those on the whiteboard, taking into consideration multiplicity but not order.

Problem 2.
The lengths of the sides of an infinite sequence of squares are 1, $\frac{1}{2}$, $\frac{1}{3}$, and so on. Determine the length of the side of the smallest square which can contain all squares in the sequence.

Problem 3.
Prove that for any real number x and any positive integer k,

$$1 - x + \frac{x^2}{2!} - \frac{x^3}{3!} + \cdots - \frac{x^{2k-1}}{(2k-1)!} + \frac{x^{2k}}{(2k)!} \geq 1.$$

1975

Problem 1.
Let a, b and c be real numbers such that $a > c \geq 0$, $b > 0$ and

$$ab^2 \left(\frac{1}{(a+c)^2} + \frac{1}{(a-c)^2} \right) = a - b.$$

Find a simpler relation among a, b and c.

Problem 2.
A quadrilateral is inscribed in a convex polygon. Is it always possible to inscribe in this polygon a rhombus whose side is not shorter than the shortest side of the quadrilateral?

Problem 3.
Let the sequence $\{x_n\}$ be defined by $x_0 = 5$ and $x_{n+1} = x_n + \frac{1}{x_n}$ for $n \geq 1$. Prove that $45 < x_{1000} < 45.1$.

1976

Problem 1.
P is a point outside a parallelogram $ABCD$ such that $\angle PAB$ and $\angle PCB$ are equal but have opposite orientations. Prove that $\angle APB = \angle DPC$.

Problem 2.
In 5^5 distinct subsets of size five of the set $\{1,2,3,\ldots,90\}$, if every two of them have a common number, prove that there exist four numbers such that each of the 5^5 subsets contains at least one of these four numbers.

Problem 3.
Prove that if $ax^2 + bx + c > 0$ for all real number x, then we can express $ax^2 + bx + c$ as the quotient of two polynomials whose coefficients are all positive.

1977

Problem 1.
Prove that for any integer $n \geq 2$, the number $n^4 + 4^n$ cannot be prime.

Problem 2.
H is the orthocenter of triangle ABC. The medians from A, B and C intersect the circumcircle at D, E and F respectively. P, Q and R are the images of D, E and F under 180° rotations about the midpoints of BC, CA and AB respectively. Prove that H lies on the circumcircle of triangle PQR.

Problem 3.
Let n be a positive integer. There are n students in each of three schools. Each student knows a total of $n + 1$ students in the other two schools. Prove that there exist three students from different schools who know one another.

1978

Problem 1.
Let a and b be rational numbers. Prove that if the equation $ax^2 + by^2 = 1$ has a rational solution (x, y), then it has infinitely many rational solutions.

Problem 2.
Let n be a positive integer. The vertices of a convex n-gon are painted so that adjacent vertices have different colors. Prove that if n is odd, then the polygon can be divided into triangles by non-intersecting diagonals such that none of these diagonals has its endpoints painted in the same color.

Problem 3.
In a triangle with no obtuse angles, r is the inradius, R is the circumradius and H is the longest altitude. Prove that we have $H \geq R + r$.

1979

Problem 1.
A convex pyramid has an odd number of lateral edges of equal length, and the dihedral angles between neighboring faces are all equal. Prove that the base is a regular polygon.

Problem 2.
A real-valued function f defined on all real numbers is such that $f(x) \leq x$ for all real number x and $f(x+y) \leq f(x)+f(y)$ for all real numbers x and y. Prove that $f(x) = x$ for all real number x.

Problem 3.
Let n be a positive integer. An $n \times n$ table of letters is such that no two rows are identical. Prove that a column may be deleted such that in the resulting $n \times n - 1$ table, no two rows are identical.

1980

Problem 1.
The points of space are painted in five colors and there is at least one point of each color. Prove that there exists a plane containing four points of different colors.

Problem 2.
Let $n > 1$ be an odd integer. Prove that there exist positive integers x and y such that

$$\frac{4}{n} = \frac{1}{x} + \frac{1}{y}$$

if and only if n has a prime divisor of the form $4k - 1$.

Problem 3.
Two tennis clubs have 1000 and 1001 members respectively. All 2001 players have different strength, and in a match between two players, the stronger one always wins. The ranking of the players within each club is known. Find a procedure using at most 11 games to determine the 1001st players in the total ranking of the 2001 players.

1981

Problem 1.
Prove that for any five points A, B, P, Q and R in a plane, $AB + PQ + QR + RP \leq AP + AQ + AR + BP + BQ + BR$.

Problem 2.
Let $n > 2$ be an even integer. The squares of an $n \times n$ chessboard are painted with $\frac{1}{2}n^2$ colors, with exactly two squares of each color. Prove that n rooks can be placed on n squares of different colors such that no two of them attack each other.

Problem 3.
For a positive integer n, $r(n)$ denote the sum of the remainders when n is divided by 1, 2, ..., n respectively. Prove that for infinitely many positive integers n, $r(n) = r(n+1)$.

1982

Problem 1.
A cube has integer side length and all four vertices of one face are lattice points. Prove that the other four vertices are also lattice points.

Problem 2.
Prove that for any integer $k > 2$, there exist infinitely many positive integers n such that the least common multiple of n, $n+1$, $\ldots$, $n+k-1$ is greater than the least common multiple of $n+1$, $n+2$, $\ldots$, $n+k$.

Problem 3.
The set of integers are painted in 100 colors, with at least one number of each color. For any two intervals $[a, b]$ and $[c, d]$ of equal length and with integral endpoints, if a has the same color as c and b has the same color as d, then $a+x$ has the same color as $c+x$ for any integer x, $0 \le x \le b-a$. Prove that the numbers 1982 and -1982 are painted in different colors.

1983

Problem 1.
Let x, y and z be rational numbers such that

$$x^3 + 3y^3 + 9z^3 - 9xyz = 0.$$

Prove that $x = y = z = 0$.

Problem 2.
In the polynomial $f(x) = x^n + a_1 x^{n-1} + \cdots + a_{n-1}x + 1$, n is a positive integer and a_1, $\ldots$, a_{n-1} are non-negative real numbers. Prove that $f(2) \ge 3^n$ if $f(x)$ has n real roots.

Problem 3.
Let n be a positive integer. P_1, P_2, $\ldots$, P_n and Q are points in the plane with no three on the same line. For any two different points P_i and P_j, there exists a third point P_k such that Q lies inside triangle $P_i P_j P_k$. Prove that n is odd.

1984

Problem 1.
When the first 4 rows of Pascal's Triangle are written down in the usual way and the numbers in vertical columns are added up, we obtain 7 numbers as shown below, 5 of them being odd.

$$
\begin{array}{ccccccc}
 & & & 1 & & & \\
 & & 1 & & 1 & & \\
 & 1 & & 2 & & 1 & \\
1 & & 3 & & 3 & & 1 \\
\hline
1 & 1 & 4 & 3 & 4 & 1 & 1
\end{array}
$$

If this procedure is applied to the first 1024 rows of Pascal's Triangle, how many of the 2047 numbers thus obtained will be odd?

Problem 2.
The rigid plates $A_1B_1A_2$, $B_1A_2B_2$, $A_2B_2A_3$, $\ldots$, $B_{13}A_{13}B_{14}$, $A_{14}B_{14}A_1$ and $B_{14}A_1B_1$ are in the shape of equilateral triangles. They can be folded along common edges A_1B_1, B_1A_2, $\ldots$, $A_{14}B_{14}$ and $B_{14}A_1$. Can they be folded so that all 28 plates lie in the same plane?

Problem 3.
Let p and q be positive integers. In a set of $n > 1$ integers, if there are two equal ones among them, add p to one and subtract q from the other. Prove that after a finite number of steps, all n numbers are distinct.

1985

Problem 1.
Let n be a positive integer. The convex $(n+1)$-gon $P_0P_1 \ldots P_n$ is divided by non-intersecting diagonals into $n-1$ triangles. Prove that these triangles can be numbered from 1 to $n-1$ such that P_i is a vertex of the triangle numbered i for $1 \le i \le n-1$.

Problem 2.
Let n be a positive integer. For each prime divisor p of n, consider the highest power of p which does not exceed n. The sum of these powers is defined as the power-sum of n. Prove that there exist infinitely many positive integers which are less than their respective power-sums.

Problem 3.
Each vertex of a triangle is reflected across the opposite side. Prove that the area of the triangle determined by the three points of reflection is less than five times the area of the original triangle.

1986

Problem 1.
Prove that three rays from the same point contain three face diagonals of some rectangular block if and only if the rays include pairwise acute angles with sum $180°$.

Problem 2.
Let n be any integer greater than 2. Determine the maximum value of h and the minimum value of H such that

$$h < \frac{a_1}{a_1 + a_2} + \frac{a_2}{a_2 + a_3} + \cdots + \frac{a_n}{a_n + a_1} < H$$

for any positive real numbers a_1, a_2, $\ldots$, a_n.

Problem 3.
From the first 100 positive integers, k of them are drawn at random and then added. For which positive integers k is the sum equally likely to be odd or even?

1987

Problem 1.
Find all positive integers a, b, c and d such that $a + b = cd$ and $ab = c + d$.

Problem 2.
Does there exist a set of points in space having at least one but finitely many points on each plane?

Problem 3.
Let n be a positive integer. Among $3n + 1$ members of a club, every two play exactly one of tennis, badminton and table tennis against each other. Each member plays each game against exactly n other members. Prove that there exist three members such that every two of them play a different game.

1988

Problem 1.
P is a point inside a convex quadrilateral $ABCD$ such that the areas of the triangles PAB, PBC, PCD and PDA are all equal. Prove that either AC or BD bisects the area of $ABCD$.

Problem 2.
From among the numbers $1, 2, \ldots, n$, we want to select triples (a, b, c) such that $a < b < c$ and, for two selected triples (a, b, c) and (a', b', c'), at most one of the equalities $a = a'$, $b = b'$ and $c = c'$ holds. What is the maximum number of such triples?

Problem 3.
The vertices of a convex quadrilateral $PQRS$ are lattice points and $\angle SPQ + \angle PQR < 180°$. T is the point of intersection of PR and QS. Prove that there exists a lattice point other than P or Q which lies inside or on the boundary of triangle PQT.

1989

Problem 1.
A circle is disjoint from a line m which is not horizontal. Construct a horizontal line such that the ratio of the lengths of the sections of this line within the circle and between m and the circle is maximum.

Problem 2.
For any positive integer m, denote by $s(m)$ the sum of its digits in base ten. For which positive integers m is it true that we have $s(mk) = s(m)$ for all integers k such that $1 \le k \le m$?

Problem 3.
From an arbitrary point (x, y) in the coordinate plane, one is allowed to move to $(x, y+2x)$, $(x, y-2x)$, $(x+2y, y)$ or $(x, x-2y)$. However, one cannot reverse the immediately preceding move. Prove that starting from the point $(1, \sqrt{2})$, it is not possible to return there after any number of moves.

1990

Problem 1.
Let p be any odd prime and n be any positive integer. Prove that at most one divisor d of pn^2 is such that $d + n^2$ is the square of an integer.

Problem 2.
I is the incenter of triangle ABC. D, E and F are the excenters opposite A, B and C respectively. The bisector of $\angle BIC$ cuts BC at P. The bisector of $\angle CIA$ cuts CA at Q. The bisector of $\angle AIB$ cuts AB at R. Prove that DP, EQ and FR are concurrent.

Problem 3.
A coin is tossed k times. Each time, the probability that it lands heads is p, where p is a real number between 0 and 1. Choose k and p for which the 2^k possible outcomes can be partitioned into 100 subsets such that the probability of the outcome being in any of the 100 subsets is the same.

1991

Problem 1.
Prove that $\dfrac{(ab+c)^n - c}{(b+c)^n - c} \leq a^n$, where n is a positive integer and $a \geq 1$, $b \geq 1$ and $c > 0$ are real numbers.

Problem 2.
A convex polyhedron has two triangular faces and three quadrilateral faces. Each vertex of one of the triangular faces is joined to the point of intersection of the diagonals of the opposite quadrilateral face. Prove that these three lines are concurrent.

Problem 3.
Given are 998 red points in the plane, no three on a line. A set of blue points is chosen so that every triangle with all three vertices among the red points contains a blue point in its interior. What is the minimum size of a set of blue points which works regardless of the positions of the red points?

1992

Problem 1.
Given n positive numbers, define their strange mean as the sum of the squares of the numbers divided by the sum of the numbers. Define their third power mean as the cube root of the arithmetic mean of their third powers. There are three mutually contradictory statements.

(1) The strange mean can never be smaller than the third power mean.

(2) The strange mean can never be larger than the third power mean.

(3) The strange mean may be larger or smaller than the third power mean.

Determine which of these statements is true for

(a) $n = 2$;

(b) $n = 3$.

Problem 2.
For an arbitrary positive integer k, let $f_1(k)$ be the square of the sum of the digits of k. For $n > 1$, let $f_n(k) = f_1(f_{n-1}(k))$. What is the value of $f_{1992}(2^{1991})$?

Problem 3.
Given a finite number of points in the plane, no three of which are collinear, prove that they can be painted in two colors so that there is no half-plane that contains exactly three given points of one color and no points of the other color.

1993

Problem 1.
Prove that if a and b are positive integers, then there are finitely many positive integers n for which both $an^2 + b$ and $a(n+1)^2 + b$ are squares of integers.

Problem 2.
The sides of triangle ABC have different lengths. Its incircle touches the sides BC, CA and AB at points K, L and M, respectively. The line parallel to LM and passing through B cuts KL at point D. The line parallel to LM and passing through C cuts MK at point E. Prove that DE passes through the midpoint of LM.

Problem 3.
Let $f(x) = x^{2n} + 2x^{2n-1} + 3x^{2n-2} + \cdots + 2nx + (2n+1)$, where n is a given positive integer. Find the minimum value of $f(x)$ for real numbers x.

1994

Problem 1.
Let λ be the ratio of the sides of a parallelogram, with $\lambda > 1$. Determine in terms of λ the maximum possible measure of the acute angle formed by the diagonals of the parallelogram.

Problem 2.
Consider the diagonals of a convex n-gon.

(a) Prove that if any $n - 3$ of them are omitted, there are $n - 3$ remaining diagonals that do not intersect inside the polygon.

(b) Prove that one can always omit $n - 2$ diagonals such that among any $n - 3$ of the remaining diagonals, there are two which intersect inside the polygon.

Problem 3.
For $1 \leq k \leq n$, the set H_k, consists of k pairwise disjoint intervals of the real line. Prove that among the intervals that form the sets H_k, one can find $\lfloor \frac{n+1}{2} \rfloor$ pairwise disjoint ones, each of which belongs to a different set H_k.

1995

Problem 1.
A lattice rectangle with sides parallel to the coordinate axes is divided into lattice triangles, each of area $\frac{1}{2}$. Prove that the number of right triangles among them is at least twice the length of the shorter side of rectangle.

Problem 2.
Each of the n variables of a polynomial is substituted with 1 or -1. If the number of -1s is even, the value of the polynomial is positive. If it is odd, the value is negative. Prove that the polynomial has a term in which the sum of the exponents of the variables is at least n.

Problem 3.
No three of the points A, B, C and D are collinear. Let E and F denote the intersection points of lines AB and CD, and of lines BC and DA, respectively. Circles are drawn with the segments AC, BD and EF as diameters. Show that either the three circles have a common point or they are pairwise disjoint.

1996

Problem 1.
In the quadrilateral $ABCD$, AC is perpendicular to BD and AB is parallel to DC. Prove that $BC \cdot DA \geq AB \cdot CD$.

Problem 2.
The same numbers of delegates from countries A and B attend a conference. Some pairs of them already know each other. Prove that there exists a non-empty set of delegates from country A such that either every delegate from country B has an even number of acquaintances among them, or every delegate from country B has an odd number of acquaintances among them.

Problem 3.
For integers $n \geq 3$ and $k \geq 0$, mark some of the diagonals of a convex n-gon. We wish to choose a polygonal line consisting of $2k + 1$ marked diagonals and not intersecting itself.

(a) Prove that this is always possible if $2kn + 1$ diagonals are marked.

(b) Prove that this may not be possible if kn diagonals are marked.

1997

Problem 1.
Let p be an odd prime number. Consider points in the coordinate plane both coordinates of which are numbers in the set $\{0, 1, 2, \ldots, p - 1\}$. Prove that it is possible to choose p of these points such that no three are collinear.

Problem 2.
The incircle of triangle ABC touches the sides at D, E and F. Prove that its circumcenter and incenter are collinear with the orthocenter of triangle DEF.

Problem 3.
Prove that the edges of a planar graph can be painted in three colors so that there is no single-color cycle.

Answers

Problem 1.

1964 The ratio $PQ : BC$ is 2:3.

1966 Five points in space with the desired properties do not exist.

1970 The number of acute angles is $2m+1$ for $n = 3m$ or $3m+1$ and $2m + 2$ for $n = 3m + 2$.

1973 We have $n = m^2 - 2$ and $k \doteq \binom{m}{2} - 1$ or $\binom{m+1}{2} - 1$ for any integer $m \geq 2$.

1975 A simpler relation is $2ab = a^2 - c^2$.

1984 There are 1365 odd numbers among the 2047.

1987 There are nine solutions: $(a, b, c, d) = (2, 2, 2, 2)$, $(1,5,2,3)$, $(5,1,2,3)$, $(1,5,3,2)$, $(5,1,3,2)$, $(2,3,1,5)$, $(3,2,1,5)$, $(2,3,5,1)$, $(3,2,5,1)$.

1992 (a) Statement (1) is true. (b) Statement (3) is true.

1994 The maximum measure of the acute angle is $\arccos \frac{\lambda^2-1}{\lambda^2+1}$.

Problem 2.

1970 The desired probability is $\frac{106081}{511038}$.

1974 The length of the side of the smallest square is $1\frac{1}{2}$.

1975 The task is not always possible.

1984 The task is not possible.

1986 The maximum value of h is 1 and the minimum value of H is $n - 1$.

1987 Such a set of points in space does exist.

1988 The maximum number of triples is $\frac{(n-1)^2}{4}$ for odd n and $\frac{n(n-2)}{4}$ for even n.

1989 The possible values are $m = 1$ and $m = 10^n - 1$ for any positive integer n.

1992 We have $f_{1992}(2^{1991}) = 256$.

Problem 3.

1966 Two infinite sets of non-negative integers with the desired properties do exist.

1971 The probability that all boxes can be opened is $\frac{1}{15}$.

1986 The sum is equally likely to be even or odd if and only if k is odd.

1990 We may have $k = 18$ and $p = 1/2 + \sqrt{\frac{4 - \sqrt[3]{10}}{12\sqrt[3]{10}}}$.

1991 The minimum size of a set of blue points is 1991.

1993 The minimum value of $f(x)$ is $n + 1$.

Part II: Discussion

Set A: Combinatorics

Combinatorics is a Hungarian specialty. It has a relatively short history compared to other disciplines within mathematics. Although it is not part of the school curriculum, it should have an essential place because it is mostly about reasoning.

An important topic in combinatorics is existence. The fundamental results are the following.

Extremal Value Principle.
In a finite set of real numbers, there exist a maximum and a minimum.

Mean Value Principle.
In a finite set of real numbers, there exist one not below average and one not above average.

Pigeonhole Principle.
There are finitely many pigeons going into finitely many holes. If there are more pigeons than holes, at least one hole will have at least two pigeons. If there are more holes than pigeons, at least one hole will have no pigeons.

Actually, only the first is a principle. The emphasis is on *finiteness*, because the real numbers can be compared two at a time. The other two are really corollaries. The Mean Value Principle follows from the observation that a maximum cannot be below average while a minimum cannot be above average.

In the Pigeonhole Principle, if there are more pigeons than holes, then the average number of pigeons per hole is greater than 1. If there are more holes than pigeons, then the average number is less than 1. No definite conclusion about existence can be drawn in case the two numbers are equal. However, there is a very important situation where there is exactly one pigeon in each hole. We then say that the pigeons and the holes are in a *one-to-one correspondence*.

25

The negation of an *existential* statement is a *universal* statement. For instance, to say that the set of positive integers does not have a maximum is to say that every positive integer is a non-maximum. Universal statements are usually easier to handle than existential statements.

Once the existence of a type of structures has been established, it is natural to ask how many of them there are, and which ones are the best accordingly to various criteria. In addition to the existence problem, we have the enumeration problem and the extremal problem. These three problems form the cornerstones of Combinatorics.

Many problems involve *probability*, such as coin-flipping and dice-rolling. When rolling a standard six-sided die, the possible outcomes are 1, 2, 3, 4, 5 and 6. They constitute the *sample space*. Suppose the desired result is rolling a prime number. Then the favorable outcomes are 2, 3 and 5. They constitute the *event*. The probability for the desired result is the size of the event divided by the size of the sample space. Thus a probability problem consists of two counting problems.

A useful problem-solving technique is using an *indirect argument*. This is sometimes called "proof by contradiction". We assume the opposite of what we are trying to prove, use correct reasoning in every step and yet arrive at a conclusion which cannot be true. The only possibility is that the original assumption is incorrect. We then have the desired result.

Another important topic in Combinatorics is *Graph Theory*. A graph is a finite set of dots and lines. The dots are called *vertices*, and various pairs of vertices are joined by lines called *edges*. An edge joining a vertex to itself is called a loop, and two edges joining the same two vertices are called multiple edges. Unless specifically permitted, we assume that there are no loops or multiple edges.

A graph is a versatile model that can represent many different situations. For instance, the vertices may represent towns and the edges may represent roads connecting them, so that the graph is a transportation network. The vertices may also represent people, and two vertices are joined by an edge if the two people they represent know each other. Then the graph becomes a social network.

There are two visual properties of graphs, connectivity and planarity. Of course, recognizing that a graph is connected or planar may not be as easy as it sounds, if the graph has a large number of vertices and edges and is drawn in a convoluted way.

A sequence $v_0 e_1 v_1 e_2 \ldots v_{n-1} e_n v_n$, where e_k is an edge joining the vertices v_{k-1} and v_k, $1 \le k \le n$, is called a *path* from v_0 to v_n. To say that a graph is *connected* means that for any two vertices, there is a path connecting them. In other words, the graph is all in one piece. If a graph comes in several pieces, it is not connected. Each piece of it is called a connected component of the graph, or simply a *component*. A closed path where $v_0 = v_n$ is called a *cycle*.

Suppose a graph has V vertices and E edges. It is called a *tree* if it has the following three properties.
(1) It is connected.
(2) It has no cycles.
(3) It satisfies the **Tree Formula:** $V = E + 1$.
These properties are not independent. Any two of them imply the third.

We first assume (1) and (2). We delete all the edges and denote by C the number of components. Since each of the V vertices is a component, we have $V = E + C$. We now restore the edges one at a time, so that E increases by 1. An edge cannot join two vertices in the same component since that will create a cycle. Hence it must join two vertices in different components, so that C decreases by 1.

It follows that we have $V = E + C$ throughout. When the last edge is restored, the graph is connected, and we have the Tree Formula.

We now assume (2) and (3). If the graph is already connected, there is nothing further to prove. Suppose it has at least two components. Adding edges joining vertices from different components will not create cycles. We continue to do so until the graph becomes connected. Then it satisfies the Tree Formula. Since it satisfies the Tree Formula to begin with, we have not really added any edges, so that the graph is already connected.

Finally, we assume (3) and (1). If the graph has no cycles, there is nothing further to prove. Suppose it has at least one cycle. Delete any edge from a cycle will not disconnect the graph. We continue to delete edges until there are no more cycles. Then the graph satisfies the Tree Formula. Since it satisfies the Tree Formula to begin with, we have not really deleted any edges, which means that the graph does not have any cycles.

From any connected graph, we can delete edges until it becomes a tree. Such a tree is called a *spanning* tree of the graph.

To say that a graph is *planar* means that the graph can be drawn in such a way that while two edges may meet at a vertex, they cannot cross each other anywhere else. A planar graph partitions the plane into non-overlapping regions. These are called the *faces* of the graph. The diagram below shows the skeleton of a cube viewed from the top. It has eight vertices, twelve edges and six faces. The infinite face is really the bottom face turned inside-out from under the other faces.

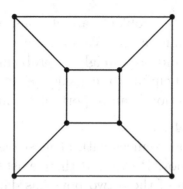

For a connected planar graph, the number V of vertices, the number E of edges, the number F of faces and the number C of connected components cannot be chosen independently. They are related by $V + F = C + E + 1$. It can be proved in the same way as the Tree Formula.

Delete all the edges. Then each vertex is a connected component so that $V = C$. Since $F = 1$ and $E = 0$, the result holds at this point. Now restore the edges one at a time, so that E goes up by 1 in each step. If the restored edge connects two vertices not connected, directly or indirectly, to each other, F does not change but C goes down by 1. If it connects two vertices already connected, F goes up by 1 but C does not change. In each case, the balance is maintained. Thus the result holds when the entire graph has been restored.

For a connected planar graph, $C = 1$ so that the above result becomes the well-known
Euler's Formula: $V - E + F = 2$.

Here is a useful corollary. Cut each edge of a planar graph into two along its length. Then we have separated the faces from one another, each bounded by at least 3 half-edges. Hence $2E \geq 3F$ so that $F \leq \frac{2E}{3}$. Substituting into Euler's Formula, we have $2 = V - E + F \leq V - E + \frac{2E}{3}$. This simplifies to $E \leq 3V - 6$.

Finally, there are problems about constructing *algorithms* to accomplish certain tasks. Well-known examples are river-crossing problems and coin-weighing problems. In such problems, it is often helpful to consider a more general case. Mathematical induction is an important technique here.

Problem A1(1964-2).

At a party, every boy dances with at least one girl, but no girl dances with every boy. Prove that there exist two boys and two girls such that each of these two boys has danced with exactly one of these two girls.

Discussion:

We are looking for two boys Ace and Cec and two girls Bea and Dee such that Ace has danced with Bea but not Dee while Cec has danced with Dee but not Bea. What are we given? First, we are told that every boy dances with at least one girl. In particular, Ace dances with some girl whom we will name Bea. Second, we are told that no girl dances with every boy. In particular, Bea does not dance with some boy whom we will name Cec. Cec in turn dances with some girl whom we will name Dee. If Dee does not dance with Ace, then we have found the boys and girls we want. All we need is to choose Ace in such a way that among the girls with whom Cec dances, there is one who does not dance with Ace.

Problem A2(1977-3).

Let n be a positive integer. There are n students in each of three schools. Each student knows a total of $n + 1$ students in the other two schools. Prove that there exist three students from different schools who know one another.

Discussion:

Note that each student knows at least one student in each of the other two schools. We seek three students Ace, Bea and Cec from different schools who know one another. How can we choose Ace so that Bea and Cec must exist?

Problem A3(1974-1).

When someone enters a library, she writes down on a blackboard the number of people already in the library at the time. When someone leaves a library, she writes down on a whiteboard the number of people still in the library. Prove that at the end of the day, the numbers on the blackboard are the same as those on the whiteboard, taking into consideration multiplicity but not order.

Discussion:

The key observation is that the identities of the persons entering or leaving the library is immaterial. Thus we can record the visits to the library during the day by a sequence of $+$s and $-$s. The sequence starts with a $+$ and ends with a $-$. Clearly, the number of $+$s is equal to the number of $-$s.

Each number on the blackboard is generated by a $+$ and each number on the whiteboard is generated by a $-$. It remains to establish a one-to-one correspondence between the $+$s and the $-$s so that the two symbols corresponding to each other generate the same number.

Problem A4(1972-2).

In a class with at least 4 students, the number of boys is equal to the number of girls. Consider all arrangements of these students in a row. Let a be the number of arrangements for which it is impossible to divide it into two parts so that the number of boys is equal to the number of girls in each part. Let b be the number of arrangement for which there is a unique partition with this property. Prove that $b = 2a$.

Discussion:

We first consider a small example with four children. Using 0 to denote one gender and 1 to denote the other gender, we have six arrangements, represented by the binary sequences 0011, 1100, 0101, 0110, 1001 and 1010.

The first two cannot be partitioned in the desired way. We say that they are of type A. In such a sequence, the running totals of 0s and 1s are never equal until the end. The remaining four sequences can be partitioned in a unique way and are said to be of type B. Such a sequence has a unique partition into two subsequences of type A. If we can establish a one-to-one correspondence between the sequences of type A and pairs of sequences of type B, then we have the desired result. Two sequences of type B form a pair if they have the same first subsequence, and the second subsequence of one is obtained from that of the other by interchanging 0s and 1s. In the example above, 0101 is paired with 0110 while 1001 is paired with 1010. Now the former will correspond to 0011 while the latter will correspond to 1100. What may be the rule?

Problem A5(1968-3).

Let $n > k > 0$ be integers. For each arrangement of n white balls and n black balls in a row, count the number of pairs of adjacent balls of different colors. Prove that the number of arrangements for which the count is $n - k$ is equal to the number of arrangements for which the count is $n + k$.

Discussion:

It may appear that we should consider establishing a one-to-one correspondence. However, a direct counting will also work. Two adjacent balls of different colors indicate a change of color if we move along the row. We can count the number of arrangements with $n + k$ changes of color and the number of arrangements with $n - k$ changes of color. Since $(n+k) + (n-k) = 2n$ is even, $n + k$ and $n - k$ are either both odd or both even.

Problem A6(1981-2).

Let $n > 2$ be an even integer. The squares of an $n \times n$ chessboard are painted with $\frac{1}{2}n^2$ colors, with exactly two squares of each color. Prove that n rooks can be placed on n squares of different colors such that no two of them attack each other.

Discussion:
A set of n squares with no two in the same row and no two in the same column is said to be an independent set. We seek an independent set with no two squares of the same color. For a particular color, how many independent sets of n squares contain both squares of that color?

Problem A7(1986-3).
From the first 100 positive integers, k of them are drawn at random and then added. For which positive integers k is the sum equally likely to be odd or even?

Discussion:
For $k = 1$, the sum consists of a single number. Since there are 50 odd numbers and 50 even numbers, the sum is equally likely to be odd or even. For $k = 2$, there are $\frac{100 \times 99}{2} = 4950$ pairs of numbers. An odd sum must be that of an odd number and an even number, and there are $50 \times 50 = 2500$ such pairs, more than half of 4950. It is reasonable to conjecture that the sum is equally likely to be odd or even for odd k, but not for even k. The concept of one-to-one correspondence may be useful.

Problem A8(1970-2).
Determine the probability that five numbers chosen at random from the first 90 positive integers contain two consecutive numbers.

Discussion:
Clearly, the number of choices of five numbers from the first 90 positive integers is $\frac{90 \times 89 \times 88 \times 87 \times 86}{5 \times 4 \times 3 \times 2 \times 1}$. We must now count the number of choices containing two consecutive numbers. It may help to count instead the number of choices in which no two numbers are consecutive.

Problem A9(1971-3).
Each of 30 boxes can be opened by a unique key. These keys are then locked at random inside the boxes, with one key in each. Two of the boxes are then broken open simultaneously, and the keys inside may be used to try to open other boxes. Keys retrieved from boxes thus opened may also be used. Determine the probability that all boxes may be opened.

Discussion:
Consider a smaller example where there are 6 box-key pairs, numbered 1 to 6. We may assume that boxes 1 and 2 are broken open. Suppose the keys inside the boxes are (3,6,1,4,2,5). We take key #3 from box #1 to open box #3, but the #1 key inside is of no further use. We use key #6 to get key #5 and then key #2. However, box #4 remains unopened. We can summarize this example more concisely by (1,3)(2,6,5)(4). Since the box-key combination is partitioned into three cycles and we only break open two boxes, the task is impossible. If we start with (3,6,1,2,4,5), then we have (1,3)(2,6,5,4) and all boxes are opened. If we start with (3,1,6,4,2,5), then we have (1,3,6,5,2)(4). Although there are only two cycles, both 1 and 2 are in the same cycle, and the task is impossible. Of course, the task is possible if there is just one cycle. Thus the problem is transformed into an equivalent one of determining when every cycle contains either 1 or 2.

Problem A10(1990-3).
A coin is tossed k times. Each time, the probability that it lands heads is p, where p is a real number between 0 and 1. Choose k and p for which the 2^k possible outcomes can be partitioned into 100 subsets such that the probability of the outcome being in any of the 100 subsets is the same.

Discussion:
We can replace 100 by smaller numbers. A fair coin with $p = \frac{1}{2}$ will work $k = 1$ for the 2 subsets {H} and {T}, and with $k = 2$ for the 4 subsets {HH}, {HT}, {TH} and {TT}. Suppose that the coin is not fair. Take $k = 2$ for the 2 subsets {HH} and {HT,TH,TT}. By the Binomial Theorem,

$$1 = (p + (1 - p))^2 = p^2 + 2p(1 - p) + (1 - p)^2.$$

Setting $\frac{1}{2} = p^2$, we have $p = \frac{1}{\sqrt{2}}$.

Problem A11(1996-2).
The same numbers of delegates from countries A and B attend a conference. Some pairs of them already know each other. Prove that there exists a non-empty set of delegates from country A such that either every delegate from country B has an even number of acquaintances among them, or every delegate from country B has an odd number of acquaintances among them.

Discussion:
Let there be n delegates from each country. Then there are $2^n - 1$ non-empty subsets of delegates from country A. Associate each with a binary n-tuple $(b_1, b_2, \ldots, b_n)$ where for $1 \le i \le n$, $b_i = 0$ if the ith delegate from country B has an even number of acquaintances among the members of the subset, and $b_i = 1$ otherwise. There are 2^n different n-tuples. If any of them is $(0,0,\ldots,0)$ or $(1,1,\ldots,1)$, there is nothing further to be proved. Assume that neither exists. Try to arrive at a contradiction.

Problem A12(1976-2).
In 5^5 distinct subsets of size five of the set $\{1,2,3,\ldots,90\}$, if every two of them have a common number, prove that there exist four numbers such that each of the 5^5 subsets contains at least one of these four numbers.

Discussion:
Use an indirect argument and assume that no such subset of four numbers exist. Let A be one of the subsets. Then each of the other $5^5 - 1$ subsets contains one of its five numbers. Suppose the number of subsets containing a particular number a in A is not less than those containing any of the other four. Then at least $5^4 + 1$ subsets contains a, including A. Now there is at least one subset B which does not contain a. Continue in this manner and try to arrive at a contradiction.

Problem A13(1987-3).
Let n be a positive integer. Among $3n + 1$ members of a club, every two play exactly one of tennis, badminton and table tennis against each other. Each member plays each game against exactly n other members. Prove that there exist three members such that every two of them play a different game.

Discussion:
We can construct a graph with $3n + 1$ vertices representing the members. Two vertices are joined by a red edge if the members they represent play tennis against each other. Similarly, green edges are used for badminton and blue edges for table tennis. Suppose there does not exist a triangle with edges with different colors. Then each triangle contains at most two pairs of adjacent sides of different colors. Try to arrive at a contradiction by counting pairs of adjacent sides of different colors, given that there are there are n edges of each color at each vertex.

Problem A14(1997-3).
Prove that the edges of a planar graph can be painted in three colors so that there is no single-color cycle.

Discussion:
A natural approach is by mathematical induction on the number of edges of the graph. The result is trivial if this number is small. In general, suppose the result holds for all planar graphs with n edges for some positive integer n.

Consider a planar graph with $n + 1$ edges. If we delete one of them, say the one joining vertex x to vertex y, we can then apply the induction hypothesis and paint the edges in three colors without creating a monochromatic cycle. If there is no monochromatic path of any one of the three colors joining x to y, we can paint the edge xy in that color. We must still make use of the assumption that the graph is planar. It places a limit on the number of edges it can have, and the existence of a monochromatic path of each color joining x to y will provide a contradiction.

Problem A15(1979-3).
Let n be a positive integer. An $n \times n$ table of letters is such that no two rows are identical. Prove that a column may be deleted such that in the resulting $n \times n - 1$ table, no two rows are identical.

Discussion:
More generally, we claim that for any $n \times k$ table with $k \geq n \geq 2$ such that no two rows are identical, we can delete $k - n + 1$ columns and leave behind an $n \times (n - 1)$ table still with no two rows identical. We use induction on n. For $n = 2$, the two rows are not identical. Hence there exists a column consisting of different letters. Deleting the other $k - 2 + 1 = k - 1$ columns will leave behind a 2×1 table with the desired property. Now try to complete the inductive argument.

Problem A16(1984-3).
Let $n > 1$, p and q be positive integers and a set of n integers. If not all n numbers are distinct, choose two equal ones, add p to one of them and subtract q from the other. Prove that after a finite number of steps, all n numbers are distinct.

Discussion:
If $n = 2$, at most 1 step will accomplish the task. Using an indirect argument, suppose there exists a $n \geq 2$ such that a finite number of steps can handle a set with n numbers, but not a set with $n + 1$ numbers. Now the largest number in the set keeps increasing while the smallest number in the set keeps decreasing. Monitor the rate of these two changes.

Problem A17(1980-3).
Two tennis clubs have 1000 and 1001 members respectively. All 2001 players have different strength, and in a match between two players, the stronger one always wins. The ranking of the players within each club is known. Find a procedure using at most 11 games to determine the 1001st players in the total ranking of the 2001 players.

Discussion:
Consider a small example with 4 and 5 members in the clubs respectively. Let the members in the first club be A_1, A_2, A_3 and A_4 with respective strengths $a_1 > a_2 > a_3 > a_4$. Let the members in the second club be B_1, B_2, B_3, B_4 and B_5 with respective strengths $b_1 > b_2 > b_3 > b_4 > b_5$. We can use a divide-and-conquer strategy, starting with a game between A_2 and B_3. Suppose $a_2 > b_3$. Then A_1 and A_2 are ahead of all of A_3, A_4, B_3, B_4 and B_5. Hence they are among the top four players. On the other hand, B_3, B_4 and B_5 are behind all of B_1, B_2, B_3, A_1 and A_2. Hence they are among the bottom five players. Suppose $a_2 < b_3$. Then B_1, B_2 and B_3 are ahead of all of A_2, A_3, A_4, B_4 and B_5. Hence they are among the top four players. On the other hand, A_2, A_3 and A_4 are behind all of A_1, B_1, B_2 and B_3. Hence they are among the bottom five players.

Set B: Number Theory

What is known as arithmetic in the school curriculum may be considered elementary Number Theory. The subject deals primarily with the properties of the positive integers.

Of the four basic operations in arithmetic, division has some implicit assumptions. Consider 3 kids dividing 6 candies. Everybody will say that the answer is $6 \div 3 = 2$. However, without the implicit assumptions, this is only one possible answer.

For example, the big kid may take 4 candies, leaving 1 each for the little ones. This highlights the first implicit assumption, that division must be *fair*. Note that fairness does not enter into addition, subtraction or multiplication. Even with this assumption, we may still have the situation where each kid gets 1 candy, saving 3 for a rainy day. Thus the second assumption is that the division must go as far as it can. We may add that the division cannot go into deficit.

With 3 kids and 8 candies, each gets 2 and 2 are saved for a rainy day. This is an instance of the following result.

Division Algorithm.
For positive integer a and d, there exists unique non-negative integers q and r such that $a = dq + r$, with $r < d$.
We call q the *quotient* and r the *remainder* of the division.

When $r = 0$, we say that a is *divisible* by d. We also say that d *divides* a, d is a *divisor* of a and a is a *multiple* of d. A useful result is that for positive integers a, b, c and d, if $a + b = c$ and d divides two of them, it will also divide the third.

Closely related to the concept of divisibility is *congruence*. Let m be a positive integer. Let a and b be two integers. They are said to be *congruent modulo m* if they both leave the same remainder when divided by m. This is denoted by $a \equiv b \pmod{m}$, and is equivalent to the statement that m is divisor of the non-negative difference of a and b.

All even integers are congruent modulo 2 to 0 while all odd integers are congruent modulo 2 to 1. The integers congruent modulo 3 to 0 are the multiples of 3. An integer congruent modulo 3 to 1 is 1 more than a multiple of 3, while an integer congruent modulo 3 to 2 is 1 less than a multiple of 3. In general, congruence modulo m partitions the set of all integers into m classes.

There are many important properties of congruence. For instance, the square of an integer is always congruent modulo 3 to 0 or 1 but never 2, and congruent modulo 4 to 0 or 1 but never 2 or 3. Every positive integer in congruent modulo 9 to the sum of its digits.

If the positive integers a and b are both divisible by a positive integer d, we say that d is a *common* divisor of a and b. The greatest among all common divisors of a and b is called their *greatest* common divisor, and is denoted by $a \triangle b$. A *common* multiple of a and b is similarly defined, but their *least* common multiple is the least *positive* integer which is a common multiple. It is denoted by $a \triangledown b$.

Like $+$, $-$, $\times$ and $\div$, $\triangle$ and $\triangledown$ are also binary operations with the following properties.

Commutative Laws.
$a \triangle b = b \triangle a$ and $a \triangledown b = b \triangledown a$.

Associative Laws.
$(a \triangle b) \triangle c = a \triangle (b \triangle c)$ and $(a \triangledown b) \triangledown c = a \triangledown (b \triangledown c)$.

De Morgan's Laws.
$a \triangle (b \triangledown c) = (a \triangle b) \triangledown (a \triangle c)$ and $a \triangledown (b \triangle c) = (a \triangledown b) \triangle (a \triangledown c)$.

Idempotent Laws.
$a \triangle a = a$ and $a \triangledown a = a$.

Product Law.
$(a \triangle b)(a \triangledown b) = ab$.

The above laws cannot be expressed so succinctly without the new notations.

The greatest common divisor can be computed by a method called the **Euclidean Algorithm**. An important consequence is that the greatest common divisor of two positive integers is a linear combination of them. In other words, there exists integers x and y such that $a \bigtriangleup b = ax + by$.

This is an example of a *Diophantine* equation, in which solutions must be integral. Many contest problems ask for the solutions of Diophantine equations. Equations with rational solutions may be converted into Diophantine equations by eliminating the common denominator.

Two positive integers are said to be *relatively prime* if their greatest common divisor is 1. Consider a product whose factors are mutually relatively prime. If a number divides this product, then it must divide one of the factors. If this product is the square of an integer, then each factor is itself the square of an integer.

The number 1 has only one positive divisor, namely, itself. Positive integers with exactly two positive divisors are called *prime* numbers. Positive integers with at least three positive divisors are called *composite* numbers. A basic result is the following.

Fundamental Theorem of Arithmetic.
Every positive integer has a unique expression as the product of a set of prime numbers in non-decreasing order.

The number 1 is the product of an empty set of prime numbers. The usual convention is that an empty sum is 0 while an empty product is 1.

Another fundamental property of the positive integers is the following result, closely related to the Extremal Value Principle.
Well-Ordering Principle.
Every non-empty subset of the set of positive integers, whether finite or infinite, has a minimum.

At first, this may appear as an easy corollary of the Extremal Value Principle, because if the set is non-empty, it must contain some positive integer, say 53. Then there are only finitely many candidates for the minimum, namely, 1, 2, 3, and so on, up to 53. By the Extremal Value Principle, this finite set has a minimum.

However, suppose the subset under consideration is actually *empty*. Then we cannot appeal to the Extremal Value Principle. The Well-Ordering Principle may be applied to show that the subset is indeed empty. Perhaps an alternative formulation of the Well-Ordering Principle is as follows. If a set of positive integers does not have a minimum, then it must be empty.

The Well-Ordering Principle is equivalent to the following two related results.

First Principle of Mathematical Induction.
Let $P(n)$ be a sequence of statements such that $P(k)$ is true for *some* positive integer k, and $P(n+1)$ is true whenever $P(n)$ is true. Then $P(n)$ is true for *all* integers $n \geq k$.

Second Principle of Mathematical Induction.
Let $P(n)$ be a sequence of statements such that $P(k)$ is true for *some* positive integer k, and $P(n+1)$ is true whenever $P(k)$, $P(k+1)$, ..., $P(n)$ are true. Then $P(n)$ is true for *all* integers $n \geq k$.

We now establish their equivalence. We first prove that the First Principle of Mathematical Induction implies the Second Principle of Mathematical Induction. Let $P(n)$ be a sequence of statements such that $P(k)$ is true, and $P(n+1)$ is true whenever $P(k)$, $P(k+1)$, ..., $P(n)$ are true. Let $Q(n)$ be the statement that $P(k)$, $P(k+1)$, ..., $P(n)$ are true. Since $P(k)$ is true, $Q(k)$ is true. Suppose $Q(n)$ is true. Then $P(k)$, $P(k+1)$, ..., $P(n)$ are true, so that $P(n+1)$ is true. It follows that $Q(n+1)$ is also true. By the First Principle of Mathematical Induction, $Q(n)$ is true for all integers $n \geq k$. This implies that $P(n)$ is also true for all integers $n \geq k$.

Next, we prove that the Second Principle of Mathematical Induction implies the Well-Ordering Principle, by showing that a subset S of the positive integers without a smallest element must be empty. Let $P(n)$ denote the statement that n is not an element of S. Now $P(1)$ must be true, as otherwise 1 is an element of S, and will be its smallest element. Suppose $P(1), P(2), \ldots, P(n)$ are all true. Then none of 1, 2, $\ldots$, n belong to S. If $n+1$ does, it will be the smallest element of S. It follows that $P(n+1)$ is also true, so that $P(n)$ is true for all positive integers n. This means that S is empty.

Finally, we prove that the Well-Ordering Principle implies the First Principle of Mathematical Induction. Let $P(n)$ be a sequence of statements such that $P(k)$ is true for some positive integer k, and $P(n+1)$ is true whenever $P(n)$ is true. Let S be the subset of the integers $n \geq k$ such that $P(n)$ is false. Suppose S is non-empty. Then it has a smallest element m. Now $m > k$ since $P(k)$ is true. Hence $m-1 \geq k$ is not in S, so that $P(m-1)$ is true. However, this implies that $P(m)$ is also true, so that m is not in S. This contradiction shows that S is empty. In other words, $P(n)$ is true for all integers $n \geq k$.

The *binomial coefficients* occur very often in counting problems. For non-negative integers n and k, $\binom{n}{l}$ is defined as the number of ways of choosing k objects from n distinct objects. Thus $\binom{4}{2} = 6$, because there are six subsets of $\{1,2,3,4\}$ of size two, namely, $\{1,2\}, \{1,3\}, \{1,4\}, \{2,3\}, \{2,4\}$ and $\{3,4\}$. Clearly, $\binom{n}{0} = 1 = \binom{n}{n}$, and $\binom{n}{k} = 0$ if $k > n$.

Note that for $n \geq k$, $\binom{n}{k} = \binom{n}{n-k}$. This is because choosing k objects we want is equivalent to choosing $n - k$ objects to discard. Similar reasoning leads to the following result.

Pascal's Formula. $\binom{n}{k} = \binom{n-1}{k-1} + \binom{n-1}{k}$.

Here is the argument. Fix a particular object. If we take it, we can choose $k - 1$ more objects from the remaining $n - 1$ objects. If we leave it, we will be taking all k objects from the remaining $n-1$ objects. Pascal's Formula follows since we either take it or leave it.

The positive integers offer many opportunities for the exploration of numerical patterns. A famous example is Pascal's Triangle.

$$\binom{0}{0}$$
$$\binom{1}{0} \qquad \binom{1}{1}$$
$$\binom{2}{0} \qquad \binom{2}{1} \qquad \binom{2}{2}$$
$$\binom{3}{0} \qquad \binom{3}{1} \qquad \binom{3}{2} \qquad \binom{3}{3}$$

All numbers on the outside are 1s by the boundary conditions $\binom{n}{0} = 1 = \binom{n}{n}$. By Pascal's formula, each number inside is the sum of the two adjacent numbers in the row above. Thus the values of the binomial coefficients are completely determined by Pascal's Triangle.

Problem B1(1987-1).
Find all positive integers a, b, c and d such that $a + b = cd$ and $ab = c + d$.

Discussion:
From $a + b + c + d = ab + cd$, we have

$$(a - 1)(b - 1) + (c - 1)(d - 1) = 2.$$

Each of $(a-1)(b-1)$ and $(c-1)(d-1)$ is a non-negative integer. Thus there are three cases to be considered.

Problem B2(1966-3).
Do there exist two infinite sets of non-negative integers such that every non-negative integer is expressible as the sum of one element from each set in a unique way?

Discussion:
We construct two sets with the desired properties. Part of every non-negative integer goes into one set while the remaining part goes into the other set. What we need is a consistent ways of doing so.

Problem B3(1969-1).
Let n be a positive integer. Prove that if $2 + 2\sqrt{28n^2 + 1}$ is an integer, then it is the square of an integer.

Discussion:
Let $m = 1 + \sqrt{28n^2 + 1}$. Then $m^2 - 2m = 28n^2$. Hence $m = 2k$ for some positive integer k, so that $7n^2 = k^2 - k = k(k - 1)$. Since k and $k - 1$ are relatively prime, one of them is a square and the other is 7 times a square. Use congruence modulo 7.

Problem B4(1982-2).
Prove that for any integer $k > 2$, there exist infinitely many positive integers n such that the least common multiple of n, $n + 1, \ldots, n + k - 1$ is greater than the least common multiple of $n + 1, n + 2, \ldots, n + k$.

Discussion:
Suppose we have $k = 3$. Take $n = 7$. Then $7 \triangledown 8 \triangledown 9 = 2^3 \times 3^2 \times 7$ while $8 \triangledown 9 \triangledown 10 = 2^3 \times 3^2 \times 5$. This value of n has the desired property. Verify that another such value is $n = 13$.

Problem B5(1990-1).
Let p be any odd prime and n be any positive integer. Prove that at most one divisor d of pn^2 is such that $d + n^2$ is the square of an integer.

Discussion:
Let d be a divisor of pn^2 such that $d + n^2 = m^2$ for some positive integer m. Try to prove that $m \triangle n = \frac{2n}{p-1}$.

Problem B6(1993-1).
Prove that if a and b are positive integers, then there are finitely many positive integers n for which both $an^2 + b$ and $a(n+1)^2 + b$ are squares of integers.

Discussion:
Let $an^2 + b = x^2$ and $a(n+1)^2 + b = y^2$ for positive integers x and y. Then $y^2 - x^2 - a = 2an$ so that

$$(y^2 - x^2 - a)^2 = 4a^2n^2 = 4a(x^2 - b).$$

Try to factor $4ab = -(y^2 - x^2)^2 + 2a(y^2 - x^2) - a^2 + 4ax^2$.

Problem B7(1985-2).
Let n be a positive integer. For each prime divisor p of n, consider the highest power of p which does not exceed n. The sum of these powers is defined as the power-sum of n. Prove that there exist infinitely many positive integers which are less than their respective power-sums.

Discussion:
Clearly, if n is a prime or a prime power, it is equal to its power-sum. Consider $6 = 2 \times 3$. For the prime 2, we have $2^2 < 6 < 2^3$. For the prime 3, we have $3 < 6 < 3^2$. Hence the power-sum of 6 is $4 + 3 = 7$, which is greater than 6 itself. We have found one number with the desired property.

Problem B8(1977-1).
Prove that for any integer $n \geq 2$, the number $n^4 + 4^n$ cannot be prime.

Discussion:
Note that n must be odd, so that $n = 2k + 1$ with $k \geq 1$. Try to factor $n^4 + 4^n = (2k + 1)^4 + 4^{2k+1}$.

Problem B9(1980-2).
Let $n > 1$ be an odd integer. Prove that there exist positive integers x and y such that $\frac{4}{n} = \frac{1}{x} + \frac{1}{y}$ if and only if n has a prime divisor of the form $4k - 1$.

Discussion:
All odd numbers are of the form $4k - 1$ or $4k + 1$. The product of an odd number of factors of the form $4k - 1$ is of the same form, while the product of an even number of factors of the form $4k - 1$ is of the form $4k + 1$. On the other hand, the product of any number of factors of the form $4k + 1$ is of the same form.

Problem B10(1978-1).
Let a and b be rational numbers. Prove that if the equation $ax^2 + by^2 = 1$ has a rational solution (x, y), then it has infinitely many rational solutions.

Discussion:
Make use of the identity $a(ux + bvy)^2 + b(vx - auy)^2 = (x^2 + aby^2)(au^2 + bv^2) = 1$ to find relations between the solutions of $ax^2 + by^2 = 1$ and the solutions of $x^2 + aby^2 = 1$.

Problem B11(1983-1).
Let x, y and z be rational numbers such that

$$x^3 + 3y^3 + 9z^3 - 9xyz = 0.$$

Prove that $x = y = z = 0$.

Discussion:
Each term of the equation is of degree 3, so that the equation is still satisfied if we cancel out the common denominator of x, y and z. Thus if there are rational solutions, there will be integer solutions. Use an indirect argument and assume that there are solutions other than $(0,0,0)$. By the Well-Ordering Principle, we can choose a solution (x, y, z) such that $|x| + |y| + |z|$ is minimum. Generate another solution in which this sum is smaller.

Problem B12(1967-1).
In a set of integers which contains both positive and negative elements, the sum of any two elements, not necessarily distinct, is also in the set. Prove that the difference between any two elements is also in the set.

Discussion:
First, prove that for any integer c in the set, nc is in the set for any positive integer n. Since the elements are all integers, the set contains a least positive integer a and a largest negative integer b by the Well-Ordering Principle. Now prove that we must have $a + b = 0$.

Problem B13(1973-1).
Determine all integers n and k, $n > k > 0$, such that the binomial coefficients $\binom{n}{k-1}$, $\binom{n}{k}$ and $\binom{n}{k+1}$ form an arithmetic progression.

Discussion:
Let $1 \leq k \leq n - 1$. We have $\binom{n}{k-1} - 2\binom{n}{k} + \binom{n}{k+1} = 0$. Simplify this expression.

Problem B14(1984-1).
When the first 4 rows of Pascal's Triangle are written down in the usual way and the numbers in vertical columns are added up, we obtain 7 numbers as shown below, 5 of them being odd.

$$
\begin{array}{ccccccc}
 & & & 1 & & & \\
 & & 1 & & 1 & & \\
 & 1 & & 2 & & 1 & \\
1 & & 3 & & 3 & & 1 \\
\hline
1 & 1 & 4 & 3 & 4 & 1 & 1 \\
\end{array}
$$

If this procedure is applied to the first 1024 rows of Pascal's Triangle, how many of the 2047 numbers thus obtained will be odd?

Discussion:
Note that $1024 = 2^{10}$. The result of taking the first $2^3 = 8$ rows is shown below. In the last row, the parity pattern of the numbers before the middle number is odd-odd-even-odd-odd-even-odd. Would this pattern hold if we take the first 2^k rows for $k \geq 4$?

```
                     ┌────┐
                     │  1 │
                     │  1 │  1
                    1│  2 │  1
                    1│  3 │  3   1
                 1  4│  6 │  4   1
                 1  5│ 10 │ 10   5   1
              1  6 15│ 20 │ 15   6   1
              1  7 21│ 35 │ 35  21   7   1
  1  1  8  7 27 20 49│ 29 │ 49  20  27   7   8   1   1
                     └────┘
```

Problem B15(1981-3).

For a positive integer n, $r(n)$ denote the sum of the remainders when n is divided by 1, 2, ..., n respectively. Prove that for infinitely many positive integers n, $r(n) = r(n+1)$.

Discussion:
For $1 \le i \le n$, let r_i denote the remainder when n is divided by i. Then $n = i\lfloor \frac{n}{i} \rfloor + r_i$. Summation yields

$$r(n) = n^2 - \sum_{i=1}^{n} i \left\lfloor \frac{n}{i} \right\rfloor.$$

Find solutions to $r(n) - r(n+1) = 0$.

Problem B16(1992-2).

For an arbitrary positive integer k, let $f_1(k)$ be the square of the sum of the digits of k. For $n > 1$, let $f_n(k) = f_1(f_{n-1}(k))$. What is the value of $f_{1992}(2^{1991})$?

Discussion:
Note that $f_1(k) \equiv k^2 \pmod 9$ for all positive integers k. Let $k = 2^{1991} \equiv 5 \pmod 9$. Find the values modulo 9 of $f_n(k)$ for $1 \le n \le 4$ and obtain upper bounds for their actual values.

Problem B17(1989-2).

For any positive integer m, denote by $s(m)$ the sum of its digits in base ten. For which positive integers m is it true that we have $s(mk) = s(m)$ for all integers k such that $1 \le k \le m$?

Discussion:
Let the set of positive integers with the desired property be M. It is easy to verify that $1 \in M$, $2 \notin M$ and $9 \in M$. For any other $m \in M$, we have $m \geq 3$ so that we can choose $k = 3$. Prove in succession that m is a multiple of 3 and of 9. Thus $m \geq 10$. By considering its base ten expression, prove that $m = 10^n - 1$ for some $n \geq 1$, and that all such numbers have the desired property.

Set C: Algebra

Algebra is the discipline which dominates the mathematics curriculum. It is very rich in routine exercises but relatively sparse in interesting problems. Thus the range of algebra problems in mathematics competitions tends to be limited. Two topics which are not covered in school play prominent parts, inequalities and functions.

The product of non-zero numbers is positive if and only if the number of negative numbers among them is even. It follows that $x^2 > 0$ for all real numbers $x \neq 0$, so that $x^2 \geq 0$ for all real numbers, with equality if and only if $x = 0$. If a sum of squares of real numbers is less than or equal to 0, then each number is equal to 0. A useful technique is the completion of squares.

Let k and $n \geq 2$ be integers and x_1, x_2, ..., x_n be positive real numbers. The kth *power mean* of these n numbers is

$$M_k = \left(\frac{x_1^k + x_2^k + \cdots + x_n^k}{n} \right)^{\frac{1}{k}}.$$

For $k = 1$, we have the *arithmetic mean* $M_1 = \frac{x_1 + x_2 + \cdots + x_n}{n}$. For $k = -1$, we have the *harmonic mean* $M_{-1} = \frac{n}{\frac{1}{x_1} + \frac{1}{x_2} + \cdots + \frac{1}{x_n}}$.

The definition does not work for $k = 0$ as $\frac{1}{k}$ is undefined, but a suitable choice is the *geometric mean* $M_0 = \sqrt[n]{x_1 x_2 \cdots x_n}$. The maximum of these n numbers may be taken as M_∞, with their minimum as $M_{-\infty}$.

An important result is the following.
Power Means Inequality.
Let $p > q$ be real numbers or $\pm\infty$. Then

$$M_p(x_1, x_2, \ldots, x_n) \geq M_q(x_1, x_2, \ldots, x_n),$$

with equality if and only if $x_1 = x_2 = \cdots = x_n$.

We prove only the special case that $M_2 \geq M_1$. This follows from

$$
\frac{x_1^2 + x_2^2 + x_3^2 + x_4^2}{4} - \left(\frac{x_1 + x_2 + x_3 + x_4}{4} \right)^2
$$

$$
= \frac{1}{16} (3(x_1^2 + x_2^2 + x_3^2 + x_4^2)
$$

$$
-2(x_1 x_2 + x_1 x_3 + x_1 x_4 + x_2 x_3 + x_2 x_4 + x_3 x_4))
$$

$$
= \frac{1}{16} ((x_1 - x_2)^2 + (x_1 - x_3)^2 + (x_1 - x_4)^2
$$

$$
+ (x_2 - x_3)^2 + (x_2 - x_4)^2 + (x_3 - x_4)^2))
$$

$$
\geq 0.
$$

A continuous function is said to be convex over an interval if for any two points in the interval, the midpoint of the chord joining two points on its graph lies above the graph. An example is x^2 on $(-\infty, \infty)$. A continuous function is said to be concave over an interval if for any two points in the interval, the midpoint of the chord joining two points on its graph lies below the graph. An example is $\sqrt{x}$ on $(0, \infty)$.

An important result about convex and concave functions is **Jensen's Inequality.**
If $f(x)$ is convex on an interval, then for any points x_1, x_2, $\ldots$, x_n in the interval,

$$
f \left(\frac{x_1 x_2 + \cdots + x_n}{n} \right) \leq \frac{f(x_1) + f(x_2) + \cdots + f(x_n)}{n}.
$$

Equality holds if and only if $x_1 = x_2 = \cdots = x_n$. The inequality is reversed if the function is concave instead.

Let P_n be the statement that if x_1, x_2, $\ldots$, x_n are not all equality, then

$$
f \left(\frac{x_1 x_2 + \cdots + x_n}{n} \right) < \frac{f(x_1) + f(x_2) + \cdots + f(x_n)}{n}.
$$

By definition, P_2 holds. Jensen's Inequality can be established by proving that P_n implies P_{n-1} as well as P_{2n}.

Apart from problems which require the proof of specific inequalities or the determination of extramal values of certain expressions, there are also problems in which inequalities are essential for their solutions.

Problem C1(1975-1).
Let a, b and c be real numbers such that $a > c \geq 0$, $b > 0$ and

$$ab^2 \left(\frac{1}{(a+c)^2} + \frac{1}{(a-c)^2} \right) = a - b.$$

Find a simpler relation among a, b and c.

Discussion:
The given equation is quadratic in b. Solving for b in terms of a and c, we can obtain a simpler relation among a, b and c.

Problem C2(1968-1).
Prove that in any infinite sequence of positive integers, it is not possible for every block of three consecutive terms a, b and c to satisfy $b = \dfrac{2ac}{a+c}$.

Discussion:
Use an indirect argument. Assume that a sequence exist such that every block of three consecutive terms a, b and c satisfy $b = \dfrac{2ac}{a+c}$. Consider the companion sequence consisting of their reciprocals.

Problem C3(1965-1).
Determine all integers a, b and c such that

$$a^2 + b^2 + c^2 + 3 \leq ab + 3b + 2c.$$

Discussion:
Since both sides are integers, we have $a^2+b^2+c^2+4 \leq ab+3b+2c$. Now complete squares.

Problem C4(1972-1).
Prove that $a(b-c)^2 + b(c-a)^2 + c(a-b)^2 + 4abc > a^3 + b^3 + c^3$,
where a, b and c are the lengths of the sides of a triangle.

Discussion:
By the Triangle Inequality, we have $b + c > a$, $c + a > b$ and
$a + b > c$. Hence $(b + c - a)(c + a - b)(a + b - c) > 0$. Show that
this equivalent to the desired inequality.

Problem C5(1991-1).
Prove that $\dfrac{(ab+c)^n - c}{(b+c)^n - c} \le a^n$, where n is a positive integer and
$a \ge 1$, $b \ge 1$ and $c > 0$ are real numbers.

Discussion:
Rewrite the inequality as $a^n c - c \le (ab + ac)^2 - (ab + c)^2$. Factor
both sides and simplify.

Problem C6(1964-3).
Prove that for any positive real numbers a, b, c and d,

$$\sqrt{\frac{a^2 + b^2 + c^2 + d^2}{4}} \ge \sqrt[3]{\frac{abc + bcd + cda + dab}{4}}.$$

Discussion:
It is sufficient to prove that

$$\sqrt[3]{\frac{abc + bcd + cda + dab}{4}} \le \frac{a + b + c + d}{4}.$$

Problem C7(1992-1).
Given n positive numbers, define their strange mean as the sum
of the squares of the numbers divided by the sum of the numbers. Define their third power mean as the cube root of the
arithmetic mean of their third powers. There are three mutually contradictory statements.

(1) The strange mean can never be smaller than the third
power mean.

(2) The strange mean can never be larger than the third power mean.

(3) The strange mean may be larger or smaller than the third power mean.

Determine which of these statements is true for

(a) $n = 2$;
(b) $n = 3$.

Discussion:
Gather numerical data. One of statements (1) or (2) will be ruled out immediately. If we find examples showing that either mean may be greater, we have established statement (3). Otherwise, it is likely that the remaining statement is true.

Problem C8(1993-3).
Let $f(x) = x^{2n} + 2x^{2n-1} + 3x^{2n-2} + \cdots + 2nx + (2n+1)$, where n is a given positive integer. Find the minimum value of $f(x)$ for real numbers x.

Discussion:
Divide $f(x)$ by $(x+1)^2 = x^2 + 2x + 1$.

Problem C9(1983-2).
In the polynomial $f(x) = x^n + a_1 x^{n-1} + \cdots + a_{n-1}x + 1$, n is a positive integer and a_1, ..., a_{n-1} are non-negative real numbers. Prove that $f(2) \geq 3^n$ if $f(x)$ has n real roots.

Discussion:
Since the coefficients of $f(x)$ are all non-negative, $f(x) > 0$ for all real numbers $x > 0$. It follows that its roots are all negative.

Problem C10(1986-2).
Let n be any integer greater than 2. Determine the maximum value of h and the minimum value of H such that
$$h < \frac{a_1}{a_1 + a_2} + \frac{a_2}{a_2 + a_3} + \cdots + \frac{a_n}{a_n + a_1} < H$$
for any positive real numbers a_1, a_2, ..., a_n.

Discussion:

Denote $\dfrac{a_1}{a_1 + a_2} + \dfrac{a_2}{a_2 + a_3} + \cdots + \dfrac{a_n}{a_n + a_1}$ by $f(a_1, a_2, \ldots, a_n)$.

Then

$$f(a_1, a_2, \ldots, a_n) + f(a_n, a_{n-1}, \ldots, a_1) = n.$$

This implies that $h + H = n$, so that we need only find h.

Problem C11(1974-3).

Prove that for any real number x and any positive integer k,

$$1 - x + \frac{x^2}{2!} - \frac{x^3}{3!} + \cdots - \frac{x^{2k-1}}{(2k-1)!} + \frac{x^{2k}}{(2k)!} \geq 1.$$

Discussion:

Let $P_{2k}(x) = 1 - x + \dfrac{x^2}{2!} - \dfrac{x^3}{3!} + \cdots - \dfrac{x^{2k-1}}{(2k-1)!} + \dfrac{x^{2k}}{(2k)!}$. Then

we have $P_{2k}(-x) = 1 + x + \dfrac{x^2}{2!} + \dfrac{x^3}{3!} + \cdots + \dfrac{x^{2k-1}}{(2k-1)!} + \dfrac{x^{2k}}{(2k)!}$,

and their product is $P_{2k}(x)P_{2k}(-x) = a_0 + a_1 x + \cdots + a_{4k}x^{4k}$.
Prove that $P_{2k}(x)P_{2k}(-x) > 0$ for all x.

Problem C12(1966-2).

Let n be any positive integer. Prove that the first n digits after
the decimal point of the decimal expansion of the real number
$(5 + \sqrt{26})^n$ are identical.

Discussion:

When $(5 + \sqrt{26})^n + (5 - \sqrt{26})^2$ is expanded, the irrational terms
cancel out, so that the expression is equal to an integer. Use the
fact that $|5 - \sqrt{26}| < \frac{1}{10}$. Consider separately the cases where n
is odd and where n is even.

Problem C13(1975-3).

Let the sequence $\{x_n\}$ be defined by $x_0 = 5$ and $x_{n+1} = x_n + \frac{1}{x_n}$ for $n \geq 1$. Prove that $45 < x_{1000} < 45.1$.

Discussion:
The lower bound is easy to prove. To establish the upper bound, note that $x_{n+1} \leq x_n^2 + 2 + \frac{1}{25} = x_n^2 + \frac{51}{25}$ since $x_n \geq 5$ holds for all $n \geq 0$. Next, show that $x_n > 15$ holds for all $n \geq 100$, so that $x_{n+1}^2 < x_n^2 + 2 + \frac{1}{225} = x_n^2 + \frac{451}{225}$.

Problem C14(1988-2).

From among the numbers 1, 2, ..., n, we want to select triples (a, b, c) such that $a < b < c$ and, for two selected triples (a, b, c) and (a', b', c'), at most one of the equalities $a = a'$, $b = b'$ and $c = c'$ holds. What is the maximum number of such triples?

Discussion:
Let us focus on (a, b, c) where b has a fixed value between 1 and n. Then the number of choices for a is $b - 1$ and the number of choices for c is $n - b$. Hence the number of choices for (a, b, c) is $\min\{b - 1, n - b\}$. Consider separately the case where n is odd and where n is even.

Problem C15(1995-2).

Each of the n variables of a polynomial is substituted with 1 or -1. If the number of -1s is even, the value of the polynomial is positive. If it is odd, the value is negative. Prove that the polynomial has a term in which the sum of the exponents of the variables is at least n.

Discussion:
We may assume that the exponent of each variable in each term is at most 1, since $1^2 = (-1)^2$. A square-free polynomial in n variables is the sum of $Ax_1 x_2 \cdots x_n + C$ plus terms of the form $Bx_{i_1} x_{i_2} \cdots x_{i_k}$, where $1 \leq k \leq n-1$. Try to prove that $A + C > 0$ and $-A + C < 0$. Then we have $A > 0$.

Problem C16(1979-2).
A real-valued function f defined on all real numbers is such that $f(x) \leq x$ for all real number x and $f(x+y) \leq f(x)+f(y)$ for all real numbers x and y. Prove that $f(x) = x$ for all real number x.

Discussion:
First prove that $f(0) = 0$. Then prove that $f(x) + f(-x) = 0$ for all real number x.

Problem C17(1976-3).
Prove that if $ax^2 + bx + c > 0$ for all real number x, then we can express $ax^2 + bx + c$ as the quotient of two polynomials whose coefficients are all positive.

Discussion:
We have $c > 0$ and $a > 0$. If $b > 0$ as well, the problem is trivial. Suppose $b < 0$. Now we must have $b^2 - 4ac < 0$ as otherwise $ax^2 + bx + c$ has real roots. Choose real numbers λ and μ such that $\lambda > -\frac{b}{c} > 0$ and $1 > \mu > \frac{b^2}{4ac} > 0$. With λ and μ fixed, choose a positive integer n such that $a + b\lambda\mu^{n-1} > 0$. Such an n exists since $\mu < 1$. Define $f(x) = \sum_{k=0}^{n} \lambda^k \mu^{\binom{k}{2}} x^k$. All its coefficients are positive. Let $g(x) = (ax^2 + bx + c)f(x)$. It remains to prove that all the coefficients of $g(x)$ are also positive.

Set D: Euclidean Geometry

Arithmetic and geometry are the oldest disciplines of mathematics, dating all the way back to ancient Greece. In fact, they were the same subject since the Greeks did arithmetic geometrically. For instance, they computed $3 + 4 = 7$ by constructing a segment of length 7 given segments of lengths 3 and 4.

The best known of the Greek geometers was undoubtedly **Euclid**, who compiled the thirteen-volume treatise titled *The Elements*. Books 7 to 9 are about arithmetic, including the Euclidean Algorithm for finding the greatest common divisor of two positive integers.

Euclidean geometry has been a mainstay in school curriculum for two thousand years. It is a fantastic subject in that there are plenty of challenging problems accessible to beginners. Many famous mathematicians attested that they were attracted to mathematics because of their exposure to and experience in Euclidean geometry.

A worrying recent trend is that Euclidean geometry is losing its prominence in the classroom worldwide. The fact that it provides challenges cuts both way. The subject is embraced by some but shunned by others, and unfortunately the latter seems to be the majority.

Mathematics competitions alleviate the situation to some extent. However, many contestants miss the point and treated the problems as obstacles to overcome rather than as opportunities for learning. They tend to tackle geometry problems by setting up coordinates and grinding things out. This simple-minded approach may work with problems involving straight lines only, but the presence of circles, especially those tangent to one another, often pose great technical difficulties.

Although most of the geometry problems ask for the proof of equalities, the first basic result is the following.

Triangle Inequality.

In a triangle, the sum of the lengths of any two sides is greater than the length of the third side.

This is a special case of a more general result that among the paths joining two given points, the segment is the shortest. Of two such paths which do not intersect, the one closer to the segment is shorter than the other. For instance, if P is a point inside triangle ABC, then $AB + AC > PB + PC$.

This can be derived from the Triangle Inequality. Extend BP to intersect at a point E. Then

$$
\begin{aligned}
AB + AC &= AB + AE + EC \\
&> BE + EC \\
&= BP + PE + EC \\
&> PB + PC.
\end{aligned}
$$

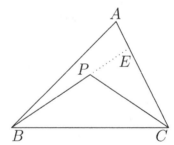

The triangle is the basic figure in Euclidean geometry. Here are some important points and lines associated with it. The line joining the midpoint of a side to the opposite vertex is called a *median*. The three medians meet at a point called the *centroid* of the triangle. The bisectors of the three angles are also concurrent, at a point called the *incenter* of the triangle. A line passing through a vertex and perpendicular to the opposite side is called an *altitude*, and the three altitudes are concurrent at a point called the *orthocenter* of the triangle. Finally, the perpendicular bisectors of the three sides are also concurrent, at a point called the *circumcenter* of the triangle.

In an equilateral triangle, all four centers coincide. Conversely, if any two of them coincide, then the triangle is equilateral. The circumcenter, the centroid and the orthocenter always lie on a line, called the *Euler* line of the triangle.

The medians, angle bisectors and altitudes of a triangle are all examples of *Cevian* lines, which join vertices to points on the opposite sides that are not themselves vertices. Note that if three Cevian lines of a triangle are concurrent, then either all three points are on the opposite sides, or only one of them, with the other two on the extensions of the sides.

A companion concept is a *Menelausian* line, which cuts the sides of a triangle at three points that are not vertices. Note that either all three points are on the extensions of the sides, or only one of them, with the other two on the sides themselves.

The concept of congruence in geometry is not to be confused with that in number theory. Two triangles ABC and DEF are said to be *congruent* if the following equalities hold.

(1) $BC = EF$ (2) $CA = FD$ (3) $AB = DE$
(4) $\angle A = \angle D$ (5) $\angle B = \angle E$ (6) $\angle C = \angle F$

Conversely, if these six equalities hold, then ABC and DEF are congruent. Actually, it is only necessary to establish three of them, provided that we choose the correct combinations. We can choose (1), (2) and (3). This is called the *SSS* case.

However, if we choose (4), (5) and (6), we have the *AAA* case which only guarantee that the two triangles have the same shape. They are congruent only if they also happen to have the same size. This is because the three angles of any triangle has a constant sum, namely, $180°$. Thus any two of (4), (5) and (6) imply the third.

Suppose we wish to choose one of (1), (2) and (3) along with two of (4), (5) and (6). Because of the angle-sum property mentioned above, we can choose any one of (1), (2) and (3) and any two of (4), (5) and (6). This is called the AAS case.

Care must be exercised if we wish to choose two of (1), (2) and (3) along with one of (4), (5) and (6). We have to make sure that the three entries come from *different* columns. Then we have congruence by the SAS case. Otherwise, we end up with the SSA case, which may or may not result in congruence.

Let ABC be a triangle in which $AB = AC$. Then we have $\angle ABC = \angle ACB$. Consider triangles BAD and CAD for any point D on BC. Since $AD = AD$, we have the SSA case, but the two triangles are congruent only if D is the midpoint of BC.

Two triangles ABC and DEF are said to be *similar* if the following equalities hold.

(1) $\frac{AB}{AC} = \frac{DE}{DF}$ (2) $\frac{BC}{BA} = \frac{EF}{ED}$ (3) $\frac{CA}{CB} = \frac{FD}{FE}$
(4) $\angle A = \angle D$ (5) $\angle B = \angle E$ (6) $\angle C = \angle F$

Conversely, if these six equalities hold, then ABC and DEF are congruent. Actually, it is only necessary to establish two of them, provided that we choose the correct combinations. We can choose any two of (1), (2) and (3). This is called the sss case. We can also choose any two of (4), (5) and (6). This is called the AA case. Finally, if we choose one of (1), (2) and (3) along with one of (4), (5) and (6), we must make sure that they are from the *same* column. This is the sAs case.

A very important special case of similarity is homothety. Two triangles ABC and APQ are said to be *homothetic* if A is the point of intersection of BP and CQ and BC is parallel to PQ. A is called the *center* of the homothety and $\frac{BC}{PQ}$ is said to be the em ratio of the homothety. We take the negative sign if A lies between B and P and the positive sign otherwise.

We denote the area of triangle ABC by $[ABC]$. The area of a triangle is equal to half the product of the length of one of its altitudes and the length of the opposite side. Here are some useful consequences of this simple formula.

Pythagoras' Theorem.
The area of the square on the hypotenuse of a right triangle is equal to the sum of areas of the squares on the two legs.

Proof:
Consider a right triangle with sides a and b and hypotenuse c. Put four copies of it around a $c \times c$ square. Since the sum of the two acute angles of the triangle is $90°$, the overall figure is a $(a+b) \times (a+b)$ square. Hence $c^2 = (a+b)^2 - 4(\frac{1}{2}ab) = a^2 + b^2$.

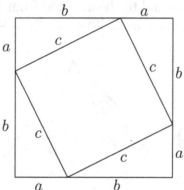

The converse of Pythagoras' Theorem may be expressed as two inequalities. If $\angle BCA < 90°$, then $AB^2 < BC^2 + CA^2$. If $\angle BCA > 90°$, then $AB^2 > BC^2 + CA^2$.

Angle Bisector Theorem.
In triangle ABC, if the bisector of $\angle A$ intersects BC at D, then $\frac{AB}{AC} = \frac{DB}{DC}$.

Proof:
Drop perpendiculars DE and DF from D to CA and AB respectively. Then triangles ADE and ADF are congruent, so that $DE = DF$. Now $[BAD] = \frac{1}{2}AB \cdot DF$ while $[CAD] = \frac{1}{2}AC \cdot DE$. Hence $\frac{AB}{AC} = \frac{[BAD]}{[CAD]} = \frac{DB}{DC}$.

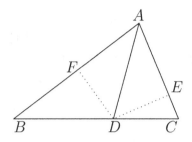

Common Side Theorem.

Triangles APQ and BPQ have a common side PQ. Let C be the point of intersection of AB and PQ. Then $\frac{[APQ]}{[BPQ]} = \frac{AC}{BC}$.

Proof:

The special case where $Q = C$ holds because the altitude from P to AQ in APQ is equal to the altitude from P to BQ in BPQ. The general case follows from $\frac{[APC]}{[BPC]} = \frac{AC}{BC} = \frac{[AQC]}{[BQC]}$.

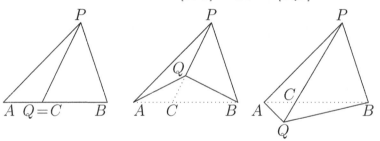

Centroid Theorem.

The medians AD, BE and CF of triangle ABC meet at the centroid G, with $\frac{AG}{GD} = \frac{BG}{GE} = \frac{CG}{GF} = 2$.

Proof:

Let BE and CF meet at G and let the extension of AG meet BC at M. By the Common Side Theorem, $\frac{[CAG]}{[BCG]} = \frac{AF}{FB} = 1$. Similarly, $[ABG] = [BCG]$ so that $\frac{BM}{MC} = \frac{[ABG]}{[CAG]} = 1$. Hence M coincides with D. Note that $[BDG] = \frac{1}{2}[BCG] = \frac{1}{2}[ABG]$. Hence $\frac{AG}{GD} = \frac{[ABG]}{[BDG]} = 2$. Similarly, $\frac{BG}{GE} = \frac{CG}{GF} = 2$.

Euler Line Theorem.

The circumcenter, the centroid and the orthocenter of a triangle are collinear.

Proof:
Let O be the circumcenter and G be the centroid of triangle ABC. Then G lies on the median AD and OD is perpendicular to BC. Extend OG to H so that $GH = 2OG$. By the Centroid Theorem, $AG = 2GD$. Hence triangles AGH and DGO are homothetic, so that AH is parallel to OD. It follows that H lies on the altitude from A. Similarly, it lies on the altitudes from B and C, so that H is indeed the orthocenter of ABC.

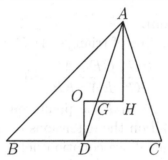

Ceva's Theorem.

If AD, BC and CF are three concurrent Cevian lines of triangle ABC, then

$$\frac{BD}{DC} \cdot \frac{CE}{EA} \cdot \frac{AF}{FB} = 1.$$

Conversely, let D, E and F be points on BC, CA and AB, respectively, either all on the sides or exactly one of them is. If the above equation holds, then AD, BC and CF are concurrent.

Proof:
Suppose the Cevian lines are concurrent at a point P. By the Common Side Theorem, we have

$$\frac{BD}{DC} \cdot \frac{CE}{EA} \cdot \frac{AF}{FB} = \frac{[PAB]}{[PAC]} \cdot \frac{[PBC]}{[PBA]} \cdot \frac{[PCA]}{[PCB]} = 1.$$

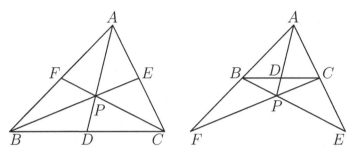

Conversely, let BE and CF intersect at P and let AP intersect BC at Q. Then $\frac{BQ}{QC} \cdot \frac{CE}{EA} \cdot \frac{AF}{FB} = 1$, so that $\frac{BD}{DC} = \frac{BQ}{QC}$. Since D and Q are either both on BC or both on its extension, we must have $D = Q$.

Menelaus' Theorem.

For any Menelausian line of triangle ABC, we have

$$\frac{BD}{DC} \cdot \frac{CE}{EA} \cdot \frac{AF}{FB} = 1.$$

Conversely, let D, E and F be points on BC, CA and AB, respectively, either all on the extensions of the sides or exactly one of them is. If the above equation holds, then D, E and F are collinear.

Proof:

By the Common Side Theorem, we have

$$\frac{BD}{DC} \cdot \frac{CE}{EA} \cdot \frac{AF}{FB} = \frac{[BEF]}{[CEF]} \cdot \frac{[CEF]}{[AEF]} \cdot \frac{[AEF]}{[BEF]} = 1.$$

Conversely, let EF intersect BC at Q. We can prove that $D = Q$ as in the proof of the converse of Ceva's Theorem.

We now consider some basic properties of circles.

Angle at the Center Theorem.
Let A, B and C be three points on a circle with center O. Then we have $\angle AOB = 2\angle ACB$.

Proof:
We have $\angle AOB = \angle OAC + \angle OCA = 2\angle ACB$ in the special case where O lies on BC. Otherwise, let PC be a diameter. By the special case, $\angle AOP = 2\angle ACP$ and $\angle BOP = 2\angle BCP$. The general case follows from these two equations.

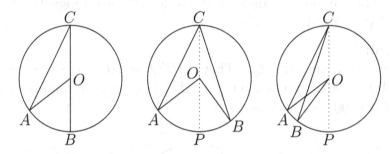

Concyclic Points Theorem.
Let A, B, C and D lie on a circle. Then
(1) $\angle ABC + \angle CDA = 180°$ and $\angle BCD + \angle DAB = 180°$;
(2) $\angle ABC = 90° = \angle CDA$ if AC is a diameter, and if BD is a diameter, then $\angle BCD = 90° = \angle DAB$.
(3) $\angle ABD = \angle ACD$, $\angle BCA = \angle BDA$, $\angle CDB = \angle CAB$ and $\angle DAC = \angle DBC$.
If any of these equations hold, A, B, C and D lie on a circle.

Proof:
Join A, B, C and D to the center O of the circle. Then we have $\angle OBA = \angle OAB$, $\angle OBC = \angle OCB$, $\angle ODC = \angle OCD$ and $\angle ODA = \angle OAD$. Summation yields (1), and (2) follows. Now take any point P on the arc BC. By (1), $\angle ABD = 180° - \angle APD = \angle ACD$. The other equations in (3) can be proved in an analogous manner. The converse result can be established by an indirect argument.

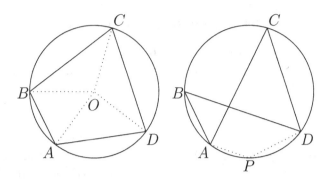

Tangent Theorem.
Let P be a point outside a circle and let PT be a tangent. Then $\angle PTA = \angle TBA$.

Proof:
Draw the diameter TQ. Then $\angle PTQ = 90°$. We also have $\angle TAQ = 90°$. Hence $\angle ATQ + \angle TQA = 90°$. It follows that $\angle PTA = 90° - \angle ATQ = \angle TBA$.

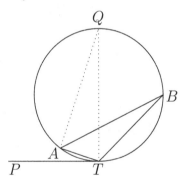

For a point P outside a circle ω with center O and radius r, the power of P with respect to ω is defined to be $OP^2 - r^2$. This is the square of the length of a tangent from P to ω.

Power of a Point Theorem.
If a line through a point P outside a circle ω intersects it at two points A and B, then $PA \cdot PB$ is equal to the power of P with respect to ω.

Proof:

Let O be the center and r the radius of ω. Let M be the midpoint of AB. Then

$$
\begin{aligned}
PA \cdot PB &= (PM - AM)(PM + AM) \\
&= (PO^2 - OM^2) - (r^2 - OM^2) \\
&= PO^2 - r^2.
\end{aligned}
$$

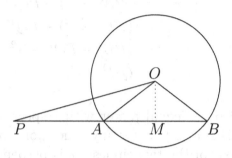

If another line through P cuts ω at two points C and D, then we have $PA \cdot PB = PC \cdot PD$. This equation also holds if two chords AB and CD intersect at a point inside the circle. Since $\angle PAC = \angle PDB$ and $\angle PCA = \angle PBD$, triangles PAC and PDB are similar. Hence $\frac{PA}{PC} = \frac{PD}{PB}$.

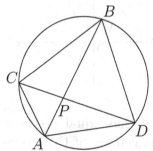

If two circles intersect at two points, then every point on the line of their common chord outside them have equal power with respect to each circle. This line is called the *radical axis* of the two circles. For two circles tangent to each other, their radical axis is the line tangent to both of them through their common point.

Suppose a point P has equal power with respect to each of two disjoint circles. Let A and B be their respective centers. Let r and s be their respective radii. Consider the line ℓ through P perpendicular to AB. It must pass between the two circles. Let it intersect AB at Q. We have

$$
\begin{aligned}
QA^2 - r^2 &= (PA^2 - PQ^2) - r^2 \\
&= (PA^2 - r^2) - PQ^2 \\
&= (PB^2 - s^2) - PQ^2 \\
&= (PB^2 - PQ^2) - s^2 \\
&= QB^2 - s^2.
\end{aligned}
$$

Hence Q also has equal power with respect to each of the two circles. This is also true for any other point on ℓ, so that ℓ is the radical axis of the two circles. It is perpendicular to the line joining the two centers, as is the case when the radical axis is a common chord or a common tangent.

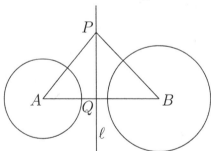

Suppose a point P has equal power with respect to each of three circles. If the centers of the circles all lie on the same line, then each pairwise radical axis is perpendicular to that line. Since P lies on all three radical axes, these axes must coincide. If the centers of the circles are not collinear, then the pairwise radical axes must be concurrent at P, making it the unique point with equal power respect to each of the three circles. It is called the *radical center* of the three circles.

There are two important circles associated with a triangle. The *circumcircle* passes through all three vertices, and its center is the circumcenter of the triangle. The *incircle* is tangent to all three sides, and its center is the incenter. Note that while the incircle is the largest circle contained by the triangle, the circumcircle is not necessarily the smallest circle containing the triangle.

An *excircle* of a triangle is tangent to one side and to the extensions of the other two sides. Its center, called an *excenter*, is the point of concurrency of the bisector of the interior angle at one vertex and the bisectors of the exterior angles at the other two vertices. Every triangle has three excircles.

Trigonometry means the geometry of the triangle. Let θ be an acute angle of a right triangle. Let h be the length of the hypotenuse, a be the length of the leg adjacent to θ and o be the length of the leg opposite to θ. The trigonometric ratios are defined as follows.

$$\sin\theta = \tfrac{o}{h} \quad \tan\theta = \tfrac{o}{a} \quad \sec\theta = \tfrac{h}{a}$$
$$\cos\theta = \tfrac{a}{h} \quad \cot\theta = \tfrac{a}{o} \quad \csc\theta = \tfrac{h}{o}$$

Here are some important trigonometric identities.

Compound Angle Formulae.

$$\sin(\alpha + \beta) = \sin\alpha\cos\beta + \cos\alpha\sin\beta,$$
$$\sin(\alpha - \beta) = \sin\alpha\cos\beta - \cos\alpha\sin\beta,$$
$$\cos(\alpha + \beta) = \cos\alpha\cos\beta - \sin\alpha\sin\beta,$$
$$\cos(\alpha - \beta) = \cos\alpha\cos\beta + \sin\alpha\sin\beta.$$

We give a geometric proof of the first identity for the special case where $\alpha + \beta < 90°$.

In the diagram below, DBE, ABD and ABC are right triangles with $\angle DBE = \alpha$ and $\angle ABD = \beta$. Then $\angle ABC = \alpha + \beta$ and we also have $\angle DAF = \alpha$. Now

$$
\begin{aligned}
\sin(\alpha + \beta) &= \frac{AC}{AB} \\
&= \frac{AF + DE}{AB} \\
&= \frac{AF}{AD} \cdot \frac{AD}{AB} + \frac{DE}{BD} \cdot \frac{BD}{AB} \\
&= \cos\alpha \sin\beta + \sin\alpha \cos\beta.
\end{aligned}
$$

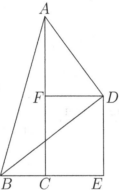

Trigonometry is a useful tool for solving certain types of geometry problems. The area of a triangle with sides a, b and c and corresponding opposite angles α, β and γ are given by $\frac{1}{2}bc\sin\alpha = \frac{1}{2}ca\sin\beta = \frac{1}{2}ab\sin\gamma$.

Here are two very important results.

Sine Formula.

Let R be the radius of the circumcircle of the triangle. Then

$$
2R = \frac{a}{\sin\alpha} = \frac{b}{\sin\beta} = \frac{c}{\sin\gamma}.
$$

Cosine Formulae.

$$
\begin{aligned}
a^2 &= b^2 + c^2 - 2bc\cos\alpha, \\
b^2 &= c^2 + a^2 + 2ca\cos\beta, \\
c^2 &= a^2 + b^2 - 2ab\cos\gamma.
\end{aligned}
$$

Finally, we prove some basic trigonometric inequalities. The main tool is Jensen's Inequality. Note that the function $\sin x$ is concave on $(0, \pi)$. In other words, for any x_1 and x_2 in $(0, \pi)$, $\frac{\sin x_1 + \sin x_2}{2} \leq \sin(\frac{x_1 + x_2}{2})$. Suppose that we have $0 < x_1 \leq x_2 < \pi$. Then we have $0 \leq \frac{x_2 - x_1}{2} < \frac{\pi}{2}$ so that $0 \leq \cos \frac{x_2 - x_1}{2} < 1$. Hence

$$\sin x_1 + \sin x_2 = 2 \sin \frac{x_1 + x_2}{2} \cos \frac{x_2 - x_1}{2} \leq 2 \sin \frac{x_1 + x_2}{2}.$$

It can be proved similarly that the function $\cos x$ is concave on $(0, \frac{\pi}{2})$ and convex on $(\frac{\pi}{2}, \pi)$.

Let A, B and C be the angles of a triangle.

Sum of Sines Inequality: $0 < \sin A + \sin B + \sin C \leq \frac{3\sqrt{3}}{2}$.
Proof:
The lower bound is trivial since $\sin x > 0$ on $(0, \pi)$. The sum approaches 0 when one angle approaches 180°. Note that $\sin x$ is concave on $(0, \pi)$. By Jensen's Inequality,

$$\sin A + \sin B + \sin C \leq 3 \sin \frac{A + B + C}{3} = \frac{3\sqrt{3}}{2}.$$

Product of Sines Inequality: $0 < \sin A \sin B \sin C \leq \frac{3\sqrt{3}}{8}$.
Proof:
The lower bound is trivial since $\sin x > 0$ on $(0, \pi)$ and the product approaches 0 when one of the angles approaches 0°. It follows from the Sum of Sines Inequality and the Arithmetic-Mean Geometric-Mean Inequality that

$$\sin A \sin B \sin C \leq \frac{\sin A + \sin B + \sin C}{3} \leq \frac{3\sqrt{3}}{8}.$$

Sum of Cosines Inequality: $1 < \cos A + \cos B + \cos C \le \frac{3}{2}$.
Proof:
We may assume that $A \le B \le C$. It follows that $B < \frac{\pi}{2}$. Note that since $\cos C = -\cos(A + B) = \sin A \sin B - \cos A \cos B$, we have $\cos A + \cos B + \cos C = 1 + \sin A \sin B - (1 - \cos A)(1 - \cos B)$. If x is an acute angle, $\sin x > \sin^2 x$ and $\cos x > \cos^2 x$ so that $\sin x + \cos x > 1$. Hence $\sin A \sin B > (1 - \cos A)(1 - \cos B)$ so that

$$\cos A + \cos B + \cos C > 1.$$

To prove the upper bound, note that if $C < \frac{\pi}{2}$, then we can use the fact that $\cos x$ is concave on $(0, \frac{\pi}{2})$. By Jensen's Inequality,

$$\cos A + \cos B + \cos C \le 3 \cos \frac{A + B + C}{3} = \frac{3}{2}.$$

If $C \ge \frac{\pi}{2}$, then $A + B \le \frac{\pi}{2}$. Since $\sin x$ is concave on $(0, \pi)$,

$$\sin A + \sin B \le 2 \sin \frac{A + B}{2} \le \sqrt{2}$$

by Jensen's Inequality. Applying the Arithmetic-Mean Geometric-Mean Inequality,

$$\sin A \sin B \le \frac{(\sin A + \sin B)^2}{4} \le \frac{1}{2}.$$

Now $(1 - \cos A)(1 - \cos B) > 0$. It follows that

$$\cos A + \cos B + \cos C < 1 + \sin A \sin B \le \frac{3}{2}.$$

Product of Cosines Inequality: $-1 < \cos A \cos B \cos C \le \frac{1}{8}$.
Proof:
The lower bound is trivial since $|\cos x| < 1$ on $(0, \pi)$. For the upper bound, we may assume that triangle ABC is acute. It then follows from the Sum of Cosines Inequality and the Arithmetic-Mean Geometric-Mean Inequality that

$$\cos A \cos B \cos C \le \frac{\cos A + \cos B + \cos C}{3} \le \frac{1}{8}.$$

Problem D1(1996-1).

In the quadrilateral $ABCD$, AC is perpendicular to BD and AB is parallel to DC. Prove that $BC \cdot DA \geq AB \cdot CD$.

Discussion:

Try to prove that $BC^2 DA^2 - AB^2 CD^2 \geq 0$.

Problem D2(1988-1).

P is a point inside a convex quadrilateral $ABCD$ such that the areas of the triangles PAB, PBC, PCD and PDA are all equal. Prove that either AC or BD bisects the area of $ABCD$.

Discussion:

Consider the two cases according to whether P lies on AC or not.

Problem D3(1981-1).

Prove that for any five points A, B, P, Q and R in a plane, $AB + PQ + QR + RP \leq AP + AQ + AR + BP + BQ + BR$.

Discussion:

Assume first that two of P, Q and R, say P and Q, are on the same side of AB while the third point, R, is on the other side. Consider the two cases where A, B, P and Q determine a quadrilateral, and where one of them lies inside the triangle determined by the other three. Use the Triangle Inequality. The argument is similar if all of P, Q and R are on the same side of AB.

Problem D4(1975-2).

A quadrilateral is inscribed in a convex polygon. Is it always possible to inscribe in this polygon a rhombus whose side is not shorter than the shortest side of the quadrilateral.

Discussion:

Investigate the special case where the overall polygon as well as the inscribed quadrilateral are both $ABCD$, a kite in which $DA = AB < BC = CD$ and $\angle BAD > \angle BCD > 120°$.

Problem D5(1967-3).
The sum of the distances from each vertex of a convex quadrilateral to the two sides not containing it is constant. Prove that the quadrilateral is a parallelogram.

Discussion:
Denote by (P, θ) the sum of the distances from a point P inside an angle θ to the arms of θ. Let P be a point inside an angle with vertex O and let the line through P perpendicular to the bisector of the angle intersect the arms of the angle at A and B. Show that for any point Q inside $\angle AOB$, $(P, \angle AOB) = (Q, \angle AOB)$ if and only if Q lies on AB.

Problem D6(1994-1).
Let λ be the ratio of the sides of a parallelogram, with $\lambda > 1$. Determine in terms of λ the maximum possible measure of the acute angle formed by the diagonals of the parallelogram.

Discussion:
Let $ABCD$ be the parallelogram with $\frac{AB}{AD} = \lambda > 1$. Let the diagonals intersect at E. Then $\angle AED < 90°$. Use the Cosine Law on triangles ADE and CBE.

Problem D7(1971-1).
A straight line intersects the sides BC, CA and AB, or their extensions, of triangle ABC at D, E and F respectively. Q and R are the images of E and F under $180°$ rotations about the midpoints of CA and AB respectively, and P is the point of intersection of BC and QR. Prove that

$$\frac{\sin EDC}{\sin RPB} = \frac{QR}{EF}.$$

Discussion:
Create lots of right triangles by constructing the altitude from A to BC and dropping perpendiculars to this line from E, F, Q and R.

Problem D8(1985-3).
Each vertex of a triangle is reflected across the opposite side. Prove that the area of the triangle determined by the three points of reflection is less than five times the area of the original triangle.

Discussion:
Let the triangle be ABC. Let $BC = a$, $CA = b$, $AB = c$, $\angle A = \alpha$, $\angle B = \beta$ and $\angle C = \gamma$. We will take $[ABC] = 1$. Let D, E and F be the respective reflectional images of A, B and C. Then $[BCD] = [CAE] = [ABF] = 1$. We have

$$[DEF] = [ABC] + [BCD] + [CAE] + [ABF]$$
$$\pm [AEF] \pm [BFD] \pm [CDE].$$

We have an acute triangle ABC in the diagram below. We take the plus sign for $[AEF]$ and the negative sign for $[BFD]$ and $[CDE]$.

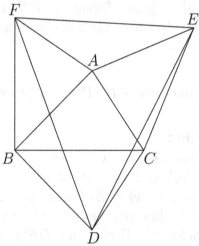

Prove that $[DEF] = 4 - \frac{1}{2}(bc \sin 3\alpha + ca \sin 3\beta + ab \sin 3\gamma)$. Then use $\sin 3\theta = \sin\theta(3 - 4\sin^2\theta)$ and $-1 < \cos\alpha\cos\beta\cos\gamma \leq \frac{1}{8}$. Remember to consider the case where ABC is an obtuse triangle.

Problem D9(1969-2).
Let the lengths of the sides of a triangle be a, b and c, and the measures of the opposite angles be α, β and γ respectively. Prove that the triangle is equilateral if

$$a(1 - 2\cos\alpha) + b(1 - 2\cos\beta) + c(1 - 2\cos\gamma) = 0.$$

Discussion:
Show that the given condition is equivalent to $\sin\alpha + \sin\beta + \sin\gamma = \sin 2\alpha + \sin 2\beta + \sin 2\gamma$. Then prove that

$$\sin\alpha + \sin\beta + \sin\gamma = 4\cos\frac{\alpha}{2}\cos\frac{\beta}{2}\cos\frac{\gamma}{2}$$

and

$$\sin 2\alpha + \sin 2\beta + \sin 2\gamma = 4\sin\alpha\sin\beta\sin\gamma.$$

Problem D10(1976-1).
P is a point outside a parallelogram $ABCD$ such that $\angle PAB$ and $\angle PCB$ are equal but have opposite orientations. Prove that $\angle APB = \angle DPC$.

Discussion:
Complete the parallelogram $BAPQ$ and prove that $CBPQ$ is cyclic.

Problem D11(1993-2).
The sides of triangle ABC have different lengths. Its incircle touches the sides BC, CA and AB at points K, L and M, respectively. The line parallel to LM and passing through B cuts KL at point D. The line parallel to LM and passing through C cuts MK at point E. Prove that DE passes through the midpoint of LM.

Discussion:
Prove that $CKEL$ and $BDKM$ are cyclic.

Problem D12(1977-2).
H is the orthocenter of triangle ABC. The medians from A, B and C intersect the circumcircle at D, E and F respectively. P, Q and R are the images of D, E and F under 180° rotations about the midpoints of BC, CA and AB respectively. Prove that H lies on the circumcircle of triangle PQR.

Discussion:
Let G be the centroid of ABC. Prove that GH is a diameter of the circumcircle of triangle PQR.

Problem D13(1995-3).
No three of the points A, B, C and D are collinear. Let E and F denote the intersection points of lines AB and CD, and of lines BC and DA, respectively. Circles are drawn with the segments AC, BD and EF as diameters. Show that either the three circles have a common point or they are pairwise disjoint.

Discussion:
First prove that the centers of the three circles are collinear. Then prove that the orthocenter of triangle ADE has equal power with respect to each of the three circles. Hence they have a common radical axis.

Problem D14(1997-2).
The incircle of triangle ABC touches the sides at D, E and F. Prove that its circumcenter and incenter are collinear with the orthocenter of triangle DEF.

Discussion:
Let O be the circumcenter and I be the incenter of triangle ABC. Let AI, BI and CI intersect the circumcircle again at L, M and N respectively. Prove that I is the orthocenter of triangle LMN so that OI is its Euler line. Let D, E and F be the points of tangency with BC, CA and AB respectively. Prove that triangle DEF is homothetic to triangle LMN.

Problem D15(1990-2).
I is the incenter of triangle ABC. D, E and F are the excenters opposite A, B and C respectively. The bisector of $\angle BIC$ cuts BC at P. The bisector of $\angle CIA$ cuts CA at Q. The bisector of $\angle AIB$ cuts AB at R. Prove that DP, EQ and FR are concurrent.

Discussion:
Let EQ and FR intersect at J, and prove that it coincides with P.

Problem D16(1989-1).
A circle is disjoint from a line m which is not horizontal. Construct a horizontal line such that the ratio of the lengths of the sections of this line within the circle and between m and the circle is maximum.

Discussion:
The special case where m is vertical is easy. Henceforth, we assume that m intersects the vertical line n through the center of the circle at some fixed point P. Show that the line joining the points of tangency from P to the circle is the desired horizontal line.

Problem D17(1978-3).
In a triangle with no obtuse angles, r is the inradius, R is the circumradius and H is the longest altitude. Prove that we have $H \geq R + r$.

Discussion:
Prove that $r = 4R \sin \frac{\alpha}{2} \sin \frac{\beta}{2} \sin \frac{\gamma}{2}$, so that $r + R$ is the sum of the distances of the circumcenter O from BC, CA and AB.

Set E: Solid Geometry & Lattice Geometry

Books 11 to 13 of *The Elements* are about *solid* geometry. The subject deals with points, lines and planes in space rather than just points and lines on a plane. Its impact on the school curriculum is practically nil in the West, and fast dwindling in the East. It seems to have vanished even from mathematics competitions. This is unfortunately because we live in a three-dimensional world. Children tend to think in concrete terms. The abstraction to two dimensions is often bewildering to them.

As an illustration, a child was given a ruler and asked to measure the space diagonal of a wooden cube. A more sophisticated person might have measured the length of a side and computed the length of the space diagonal via Pythagoras' Theorem. Instead, the child placed the base $ABCD$ of the cube on a rectangular table, with A on a corner and AB and AD along two sides of the table. The cube was then tipped over BC to create a virtual cube of the same size. The length of its space diagonal could be measured directly from the corner of the table to D.

It is ironic that sometimes, a result in plane geometry is harder to prove than its analog in space. Such an example is **Desargues' Theorem.**
Let ABC and $A'B'C'$ be two triangles. Suppose AA', BB' and CC' are concurrent at some point V, BC and $B'C'$ meet at L, CA and $C'A'$ meet at M, and AB and $A'B'$ meet at N. Then L, M and N are collinear.

This result has a converse. Let ABC and $A'B'C'$ be two triangles. Suppose BC and $B'C'$ meet at L, CA and $C'A'$ meet at M, AB and $A'B'$ meet at N, where L, M and N are collinear. Then AA', BB' and CC' are concurrent.

81

The converse actually follows from Desargues' Theorem itself. Consider triangles $BB'N$ and $CC'M$. Note that $B'N$ cuts $C'M$ at D while NB cuts MC at A. Since BC cuts $B'C'$ at L, the lines BB' and CC' are coplanar, and will meet at some point V. Since $BC, B'C'$ and MN are concurrent at L, it follows from Desargues' Theorem that V, A and A' are collinear, which is equivalent to the desired result.

In the plane, two lines either intersect or are parallel. In space, there is a third possibility. Two lines which do not intersect and are not parallel are called *skew* lines, with respect to each other, such as a straight bridge over a straight river.

Two planes are either parallel or intersect along a line. For three planes, they may all be parallel. We may have two parallel planes while the third one intersect each of these two in a line. We may have all three planes intersect in the same line or in three parallel lines. Finally, the three planes may have just a single common point, such as the origin in each of the three coordinate planes.

When two planes intersect, they determine an angle between them called a *dihedral angle*. Let ℓ be the line of intersection of the two planes and let P be any point on ℓ. Let Q be a point on one plane and R be a point on the other plane such that PQ and PR are perpendicular to ℓ. Then the dihedral angle is equal to $\angle QPR$.

We now present and prove the space analog of Desargues' Theorem.

Desargues' Theorem in Space.
Let Π and Π' be two planes and V be a point not in either of them. Let triangle ABC in Π be projected from V onto triangle $A'B'C'$ in Π'. Then the lines BC and $B'C'$ will intersect, as will CA and $C'A'$, as well as AB and $A'B'$. Moreover, these three points of intersection are collinear.

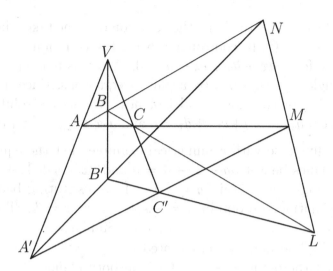

Proof:

Note that B' and C' lie on the plane VBB', so that the coplanar lines BC and $B'C'$ will intersect at some point L. Similarly, CA cuts $C'A'$ at some point M, and AB cuts $A'B'$ at some point N. Now Π and Π' meet along a line ℓ. Since L lies in both Π and Π', it lies on ℓ. Similarly, so do M and N. Hence these three points are collinear.

We now derive Desargues' Theorem from its space analog. Through V, draw a line not on the plane Π of ABC. Take on this line points D and D'. Now BCD and $B'C'D'$ are two non-coplanar triangles satisfying the hypothesis of Desargues' Theorem in Space. Hence BD meets $B'D'$ at some point Y, and CD meets $C'D'$ at some point Z. Moreover, L, Y and Z are collinear. Similarly, AD meets $A'D'$ at some point X, and M, Z and X are collinear, as are N, X and Y. Note that none of X, Y and Z is in Π. A plane Γ containing X, Y and Z will also contain L, M and N, and L, M and N all lie on the line of intersection of Π and Γ.

In the coordinate plane, the equation of a line takes the form $ax + by + c = 0$. In coordinate space, the equation of a plane takes the form $ax + by + cz + d = 0$. A line is not represented by a single equation but by a pair of equations, since it is the intersection of two non-parallel planes. We seek a solution to the system $a_1 x + b_1 y + c_1 z + d_1 = 0$ and $a_2 x + b_2 y + c_2 z + d_2 = 0$.

We illustrate with a numerical example. Let the equations of the planes be $x + 3y - 2z + 1 = 0$ and $2x - y + 3z + 2 = 0$. Eliminating z, we obtain $y = -x - 1$. Substituting back into either equation, we obtain $z = -x - 1$. Let $x = k$. Then we have $y = -k - 1$ so that $k = -(y + 1)$. Similarly, $k = -(z + 1)$. Hence the line may be represented as $\frac{x-0}{1} = \frac{y-(-1)}{-1} = \frac{z-(-1)}{-1}$. Note that the point $(0, -1, -1)$ lies on both planes.

In general, a line may be represented as $\frac{x-a}{\ell} = \frac{y-b}{m} = \frac{z-c}{n}$, where (a, b, c) is a point on the line, and the numbers ℓ, m and n give the direction of the line. This is the three-dimensional analog of the point-slope formula in the plane.

The *vector* going from the point $A(a_1, a_2, a_3)$ to the point $B(b_1, b_2, b_3)$ is denoted by $\mathbf{AB} = \mathbf{v} = (v_1, v_2, v_3)$, where $v_1 = b_1 - a_1$, $v_2 = b_2 - a_2$ and $v_3 = b_3 - a_3$. Thus (ℓ, m, n) is the vector of the line $\frac{x-a}{\ell} = \frac{y-b}{m} = \frac{z-c}{n}$.

A vector $\mathbf{v}$ has a length, denoted by $|\mathbf{v}|$, as well as a direction. Two vectors with the same length and the same direction are considered the same, even though they may have different initial points, and consequently different terminal points. Thus vectors can be translated freely.

The *scalar multiple* of a vector $\mathbf{v}$, with a real number λ as the multiplier, is the vector $\lambda \mathbf{v}$ which is in the same direction as $\mathbf{v}$ but λ times its length. The term scalar is used in contrast to the term vector. It is just a real number and has no directions.

The *sum* of two vectors is defined as follows. If the vectors are in the same direction, their sum is a vector in the same direction, and its length is the sum of the lengths of the individual vector. The sum of two vectors in opposite directions is a vector in the direction of the longer vector, and its length is the difference of the lengths of the individual vectors. Of course, if the individual lengths are the same, then the *zero* vector results.

Let $\mathbf{u}$ and $\mathbf{v}$ be two non-collinear vectors. Let $\mathbf{u}$ go from $A(a_1, a_2, a_3)$ to $B(b_1, b_2, b_3)$ while $\mathbf{v}$ goes from A to $C(c_1, c_2, c_3)$. Translate $\mathbf{v} = (c_1 - a_1, c_2 - a_2, c_3 - a_3)$ to $\mathbf{BP}$. Then P is the point $(p_1, p_2, p_3) = (b_1 + c_1 - a_1, b_2 + c_2 - a_2, b_3 + c_3 - a_3)$. Hence $p_1 + a_1 = b_1 + c_1$, $p_2 + a_2 = b_2 + c_2$ and $p_3 + a_3 = b_3 + c_3$, so that $ABPC$ is a parallelogram. We define $\mathbf{u} + \mathbf{v}$ as the vector $\mathbf{AP}$.

For two vectors $\mathbf{u} = (u_1, u_2, u_3)$ and $\mathbf{v} = (v_1, v_2, v_3)$, their *dot product* is a scalar defined by $\mathbf{u} \cdot \mathbf{v} = |\mathbf{u}||\mathbf{v}| \cos \theta$, where θ is the angle between the vectors. It follows from the Cosine Formula that $\cos \theta = \frac{u_1^2 + u_2^2 + u_3^2 + v_1^2 + v_2^2 + v_3^2 - (v_1 - u_1)^2 - (v_2 - u_2)^2 - (v_3 - u_3)^2}{2\sqrt{u_1^2 + u_2^2 + u_3^2}\sqrt{v_1^2 + v_2^2 + v_3^2}}$. Hence $\mathbf{u} \cdot \mathbf{v} = u_1 v_1 + u_2 v_2 + u_3 v_3$. Note that two vectors are perpendicular if their dot product is 0.

Let $\mathbf{u} = (u_1, u_2, u_3)$ and $\mathbf{v} = (v_1, v_2, v_3)$ be two non-parallel vectors. Their *cross product* is a vector $\mathbf{u} \times \mathbf{v} = \mathbf{w} = (w_1, w_2, w_3)$ which is perpendicular to both of them. Eliminating w_3 from $u_1 w_1 + u_2 w_2 + u_3 w_3 = 0$ and $v_1 w_1 + v_2 w_2 + v_3 w_3 = 0$, we have $w_1(u_1 v_3 - v u_3 v_1) + w_2(u_2 v_3 - u_3 v_2) = 0$. Setting $w_1 = u_2 v_3 - u_3 v_2$, we have $w_2 = u_3 v_1 - u_1 v_3$. Hence $w_3 = u_1 v_2 - u_2 v_1$.

The distance between two intersecting planes is 0. The distance between two parallel planes is the length of a segment intersecting them at right angles. Similarly, the distance between two intersecting lines is 0. The distance between two parallel lines is the length of a segment intersecting both of them at right angles. This segment lies on the plane determined by the two lines.

Two skew lines ℓ_1 and ℓ_2 do not determine a plane, but the distance between them is still the length of a segment intersecting both of them at right angles. Such a segment always exists. We may assume that ℓ_1 lies on a horizontal plane Π_1. Rotate space about ℓ_1 until ℓ_2 also lies on a horizontal plane Π_2. The distance between ℓ_1 and ℓ_2 will be the same as the distance between Π_1 and Π_2.

We now construct the common perpendicular of ℓ_1 and ℓ_2. Project ℓ_2 onto Π_1. Its image will intersect ℓ_1 at a point P_1. This is one endpoint of the common perpendicular. The other endpoint P_2 is obtained by projecting P_1 onto Π_2. Note that the direction of the common perpendicular is given by the cross product of the vectors of the two lines.

In *lattice* geometry, the integer points on the number line constitute a one-dimensional lattice. The points (x, y) in the plane where both coordinates x and y are integers constitute a two-dimensional lattice. The points (x, y, z) in space where all three coordinates x, y and z are integers constitute a three-dimensional lattice.

A well-known problem is to prove that among any five lattice points in the plane, there exist two of them such that the line segment joining them passes through another lattice points. The proof uses parity and the Pigeonhole Principle.

The lattice points (x, y) may be divided into four classes, according to whether each of x and y is even or odd. Since there are five points, two of them must be in the same class, say (x_1, y_1) and (x_2, y_2). Since x_1 and x_2 have the same parity, $\frac{x_1+x_2}{2}$ is an integer. Similarly, so is $\frac{y_1+y_2}{2}$. Hence $\left(\frac{x_1+x+2}{2}, \frac{y_1+y_2}{2}\right)$ is a lattice point, and it is the midpoint of the segment joining (x_1, y_1) and (x_2, y_2).

A lattice polygon is one in which all vertices are lattice points in the plane. The diagram below shows an example with seven lattice points on its boundary and ten lattice points in its interior.

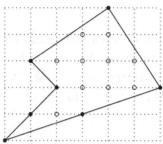

Perhaps the most well-known result in lattice geometry is **Pick's Formula.**

Let I and B be the numbers of points in the interior and on the boundary, respectively, of a lattice polygon. Then the area of the polygon is $I + \frac{B}{2} - 1$.

Pick's Formula can be proved using mathematical induction on the number n of sides of the lattice polygon. For $n = 3$, consider the smallest rectangle R which contains a lattice triangle T. The sides of R sides are parallel to the coordinate axes.

Each side of R must contain at least one vertex of T. The diagram below shows all possible cases.

(1) All three vertices of T are vertices of R.

(2) Two vertices of T are adjacent vertices of R.

(3) Two vertices of T are opposite vertices of R.

(4) Only one vertex of T is a vertex of R.

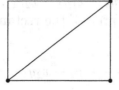 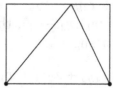

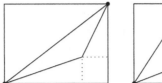

Pick's Formula can be proved to hold in each of these cases. This is one of the rare examples in mathematical induction in which the base step requires more work. We will only deal with the last case. The others can be handled in essentially the same manner.

The diagram below shows a lattice triangle with vertices (x_1, y_1), (x_2, y_2) and (x_3, y_3), and the smallest rectangle which contains it. The sides of the rectangle are parallel to the coordinate axes. The numbers of lattice points on the three sides of the triangle, excluding the vertices, are denoted by a, b and c.

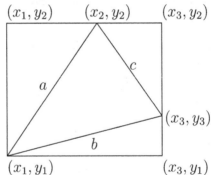

The areas of the corner triangles are $\frac{1}{2}(x_2 - x_1)(y_2 - y_1)$, $\frac{1}{2}(x_3 - x_1)(y_3 - y_1)$ and $\frac{1}{2}(x_3 - x_2)(y_2 - y_3)$. Their total area is $x_1 y_1 + \frac{1}{2}(x_2 y_3 + x_3 y_2 - x_3 y_1 - x_1 y_3 - x_1 y_2 - x_2 y_1)$. Subtracting this from $(x_3 - x_1)(y_2 - y_1)$, the area of the rectangle, the area of the central triangle is

$$S = \frac{1}{2}(x_1 y_3 + x_2 y_1 + x_3 y_2 - x_1 y_2 - x_2 y_3 - x_3 y_1).$$

For the central triangle, $B = 3 + a + b + c$. The number of lattice points inside the rectangle is $(x_3 - x_1 - 1)(y_2 - y_1 - 1)$. The number of lattice points inside the corner triangle which has a lattice points on one of its sides is $\frac{1}{2}((x_2 - x_1 - 1)(y_2 - y_1 - 1) - a)$. Similarly, we can prove that the other two corner triangles have $\frac{1}{2}((x_3 - x_1 - 1)(y_3 - y_1 - 1) - b)$ and $\frac{1}{2}((x_3 - x_2 - 1)(y_2 - y_3 - 1) - c)$ lattice points inside them, respectively. It follows that we have $I = S - \frac{1+a+b+c}{2}$ so that $S = I + \frac{B}{2} - 1$.

The inductive step is easy. For $n \geq 4$, use an interior diagonal to divide the n-gon into two polygons each with less than n sides. Let the numbers of lattice points in their interiors be I_1 and I_2 respectively. Let the numbers of lattice points on their boundaries be B_1 and B_2 respectively. Let k be the number of lattice points inside the n-gon which lies on the dividing diagonal. Then $I = I_1 + I_2 + k$ and $B = B_1 + B_2 - (2k + 2)$. The area of the n-gon is

$$I_1 + I_2 + \frac{B_1 + B_2}{2} - 2 = I - k + \frac{B + 2k + 2}{2} - 2 = I + \frac{B}{2} - 1.$$

The lattices defined so far are called *orthogonal* lattices. There are other kinds such as an *isogonal* lattice. It is well-known that there are three regular polygons which can tessellate the plane. The tessellation with squares is the basis of the orthogonal lattice. If the lattice points are defined instead to be the vertices in the tessellation with equilateral triangles, we have the isogonal lattice. The vertices in the tessellations with regular hexagons form a subset of the isogonal lattice.

Problem E1(1969-3).
A $1 \times 8 \times 8$ block consists of 64 unit cubes such that exactly one face of each cube is painted black. The initial arrangement is arbitrary. In a move, we may rotate a row or a column of 8 cubes about their common axis. Prove that after a finite number of such moves, we can obtain an arrangement in which all the black faces are on top.

Discussion:
Let us consider a smaller problem with a $1 \times 2 \times 2$ block. It consists of just 4 unit cubes, arranged in rows 1 and 2 as well as column a and b. A cube is said to face in the direction of its black face. Each cube may be facing front, back, left, right, top or bottom. We first make the cube a1 face front. If it is facing back, top or bottom, rotate row 1 so that it faces front. If it is facing left or right, we first rotate column a until a1 faces top or bottom before rotating row 1. Then further rotation of column a leaves a1 facing front. In the same way, we can make a2 face front. Now we rotate the rows to make column a face top. This is followed by rotating column a to make column a face left. Then further rotations of the rows leave column a facing left. In the same way, we can make column b face left. When all the cubes are facing left, rotations of the columns can make them all face top.

Problem E2(1980-1).
The points of space are painted in five colors and there is at least one point of each color. Prove that there exists a plane containing four points of different colors.

Discussion:
Suppose A, B, C, D and E are five points of different colors. If any four of them are coplanar, there is nothing further to prove. If $ABCD$ is a tetrahedron, see whether or not the line AE intersects the plane BCD.

Problem E3(1987-2).
Does there exist a set of points in space having at least one but finitely many points on each plane?

Discussion:
The equation of a plane takes the form $ax + by + cz + d = 0$. Express each of x, y and z in terms of a parameter t so that we have a polynomial equation in t. We want this equation to have at least one but finitely many real roots.

Problem E4(1966-1).

Do there exist five points A, B, C, D and E in space such that

$$AB = BC = CD = DE = EA \quad \text{and}$$
$$\angle ABC = \angle BCD = \angle CDE = \angle DEA = \angle EAB = 90°?$$

Discussion:

Suppose such a pentagon $ABCDE$ exists. We may take the length of each side to be 2. Then the length of each diagonal is $2\sqrt{2}$. Place the right isosceles triangle BCD on a horizontal plane. Where can the points A and E possibly be?

Problem E5(1973-3).

Let n be an integer greater than 4. Every three of n planes have a common point, but no four of these planes have a common point. Prove that among the regions into which space is divided by these planes, the number of tetrahedra is not less than $\frac{2n-3}{4}$.

Discussion:

Call a point of intersection of three of the n planes a vertex. Let P be a vertex on one side of a plane Π_0 and closest to Π_0. Let P be determined by Π_1, Π_2 and Π_3. These 4 planes determine a tetrahedron $PQRS$ with Q, R and S on Π_0. Prove that $PQRS$ is among the tetrahedra we seek, and we say that it is associated with Π_0. Count how many tetrahedra are associated with a plane, and how many planes are associated with a tetrahedron.

Problem E6(1991-2).

A convex polyhedron has two triangular faces and three quadrilateral faces. Each vertex of one of the triangular faces is joined to the point of intersection of the diagonals of the opposite quadrilateral face. Prove that these three lines are concurrent.

Discussion:

Let the triangular faces be ABC and DEF, connected by the edges AD, BE and CF. No two of the planes BCD, CAE and ABF are parallel. Hence they have a unique common point. Prove that it is the desired point of concurrency.

Problem E7(1979-1).
A convex pyramid has an odd number of lateral edges of equal length, and the dihedral angles between neighboring faces are all equal. Prove that the base is a regular polygon.

Discussion:
Prove that the foot of the perpendicular from the apex of the pyramid to the base is equidistant from every vertex of the base, so that the base has a circumcircle. Prove also that every other side of the base has the same length.

Problem E8(1964-1).
$PABC$ is a triangular pyramid with $AB = BC = CA$ and $PA = PB = PC$. Another triangular pyramid congruent to $PABC$ is glued to it along the common base ABC to obtain a hexahedron with five vertices such that the dihedral angle between any two adjacent faces is the same. Determine the ratio $PQ : BC$ where Q is the fifth vertex.

Discussion:
Such a hexahedron may be obtained by rotating an equilateral triangle APQ 120° clockwise about PQ and then 120° counterclockwise about PQ. Let B and C be the images of A. Verify that the hexahedron $P - ABC - Q$ has the desired property, and that it is the only hexahedron which has this property.

Problem E9(1986-1).
Prove that three rays from the same point contain three face diagonals of some rectangular block if and only if the rays include pairwise acute angles with sum 180°.

Discussion:
For necessity, consider any rectangular block $ABCD - EFGH$ and focus on the face diagonals AC, AF and AH. Prove that $\angle CAF + \angle CAH + \angle FAH = 180°$, and that these three angles are acute. For sufficiency, try to construct a rectangular block with the desired properties.

Problem E10(1965-3).

The base $ABCD$ and the top $EFGH$ of a hexahedron are both squares. The lateral edges AE, BF, CG and DH have equal length. The circumradius of $EFGH$ is less than the circumradius of $ABCD$, which is in turn less than the circumradius of $ABFE$. Prove that the shortest path on the surface of this hexahedron going from A to G passes through only the lateral faces.

Discussion:

By symmetry, there are three possible types of paths from A to G, crossing CD, EF and BF respectively. In each case, we unfold the surface of the hexahedron along the specified edge and connect A and G by a line segment. In the first case, AG clearly intersects CD. In the second case, if AG misses EF, then it intersects EH so that $45° > \angle HGA = \angle GAB > \angle EAB$, which is a contradiction. In the third case, AG will intersect BF since $\angle AFG = \angle AFB + \angle BFG < 45° + 135° = 180°$. Now compare the lengths of the segment AG in the three cases.

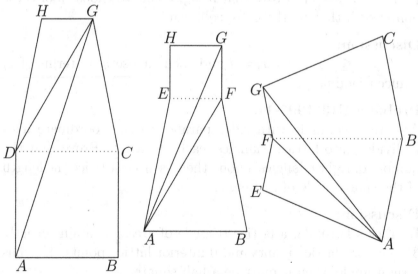

Problem E11(1982-3).
The set of integers are painted in 100 colors, with at least one number of each color. For any two intervals $[a, b]$ and $[c, d]$ of equal length and with integral endpoints, if a has the same color as c and b has the same color as d, then $a + x$ has the same color as $c + x$ for any integer x, $0 \leq x \leq b - a$. Prove that the numbers 1982 and -1982 are painted in different colors.

Discussion:
We first consider a particular coloring scheme. Let the colors be 1 to 100. An integer is painted in color k if it is congruent modulo 100 to k. This satisfies the hypothesis of the problem. Since -1982 and 1982 are not congruent modulo 100, they are painted in different colors in this scheme. Now prove that this scheme is the only possible one. Use the Pigeonhole Principle.

Problem E12(1997-1).
Let p be an odd prime number. Consider points in the coordinate plane both coordinates of which are numbers in the set $\{0, 1, 2, \ldots, p - 1\}$. Prove that it is possible to choose p of these points such that no three are collinear.

Discussion:
For $0 \leq i \leq p - 1$, let $x_i = i$ and then choose y_i in terms of x_i reduced modulo p.

Problem E13(1995-1).
A lattice rectangle with sides parallel to the coordinate axes is divided into lattice triangles, each of area $\frac{1}{2}$. Prove that the number of right triangles among them is at least twice the length of the shorter side of rectangle.

Discussion:
By Pick's Formula, a lattice triangle of area $\frac{1}{2}$ contains exactly 3 boundary lattice points and 0 interior lattice points. If it has a right angle, then it must be a half-square.

Two half-squares can be put together to form a parallelogram with sides 1 and $\sqrt{2}$. These are called basic parallelograms. Prove that the triangulation may be modified so that the triangles which are not half-squares can be combined into basic parallelograms.

Problem E14(1988-3).
The vertices of a convex quadrilateral $PQRS$ are lattice points and $\angle SPQ + \angle PQR < 180°$. T is the point of intersection of PR and QS. Prove that there exists a lattice point other than P or Q which lies inside or on the boundary of triangle PQT.

Discussion:
By symmetry, we may assume that $\angle PQR + \angle QRS \leq 180°$. Let S be closer to QR than P. Construct the parallelogram $PSRR'$. Since P, S and R are lattice points, so is R'. Note that R' is inside triangle PQR but cannot coincide with P or Q. If it is inside or on the boundary of triangle PQT, there is nothing further to prove. What happens if R' is in the interior of triangle QRT?

Problem E15(1973-2).
For any point on the plane of a circle other than its center, the line through the point and the center intersects the circle at two points. The distance from the point to the nearer intersection point is defined as the distance from the point to the circle. Prove that for any positive number ϵ, there exists a lattice point whose distance from a given circle with center $(0,0)$ and radius r is less than ϵ, provided that r is sufficiently large.

Discussion:
Consider the part of the circle in the first quadrant. Choose a point P on the x-axis such that the vertical line through P intersects the circle at a point C which is not a lattice point, and $OP > r - 1$, where O is the center of the circle. Show that the first lattice point directly below C is the point we seek.

Problem E16(1982-1).

A cube has integer side length and all four vertices of one face are lattice points. Prove that the other four vertices are also lattice points.

Discussion:

Let the cube be $ABCD - EFGH$ with edges AE, BF, CG and DH between the bases $ABCD$ and $EFGH$. Let A, B, C and D be lattice points. We may take A to be the origin, and the let the coordinates for B, D and E be (b_1, b_2, b_3), (d_1, d_2, d_3) and (e_1, e_2, e_3) respectively, where b_1, b_2, b_3, d_1, d_2 and d_3 are integers. The desired result will follow if we can show that e_1, e_2 and e_3 are also integers. Use vectors.

Problem E17(1984-2).

The rigid plates $A_1B_1A_2$, $B_1A_2B_2$, $A_2B_2A_3$, ..., $B_{13}A_{13}B_{14}$, $A_{14}B_{14}A_1$ and $B_{14}A_1B_1$ are in the shape of equilateral triangles such that they can be folded along common edges A_1B_1, B_1A_2, ..., $A_{14}B_{14}$ and $B_{14}A_1$. Can they be folded so that all 28 plates lie in the same plane?

Discussion:

Use an indirect argument and assume that the plates can lie in the same plane. Mark off an isogonal lattice so that when a plate is placed in the plane, its three vertices are lattice points. Paint the lattice points in three colors so that when a plate is placed on the lattice, all three vertices are lattice points of different colors. Try to find a contradiction.

Set F: Combinatorial Geometry

Combinatorial geometry is one of the newest disciplines in mathematics. As its name suggests, it deals with combinatorial problems with geometrical settings. It has become a very important topic in mathematics competitions, because refreshing problems are increasingly difficult to find in other traditional disciplines.

Unfortunately, the topic is still in its formative stage. Each problem has its own individual characteristics, which makes it difficult to conduct a comprehensive general discussion. It may be possible to do so some day, but in the meantime, we simply have to learn as we go along.

Some problems deal with sets of points and lines in the plane. A set of points is said to be in *general* position if no three of them are collinear. A set of lines is said to be in *general* position if no three of them are concurrent and no two of them are parallel.

A well-known problem asks for the proof that given an even number of points in general position, the points can be joined in pairs by non-intersecting segments.

Since the number of points is finite, the number of configurations in which the points are joined in pairs by segment, intersecting or otherwise, is also finite. By the Extremal Value Principle, there exists a configuration in which the total length of the segments is minimum.

We use an indirect argument to prove that this configuration has the desired property. Suppose to the contrary that the points A, B, C and D are joined by segments AB and CD intersecting at E, as shown in the diagram below. By the Triangle Inequality, $AB + CD = (AE + CE) + (BE + DE) > AC + BD$. If we use AC and BD to join the four points instead of AB and CD, we have reduced the total length of the segments. This is a contradiction.

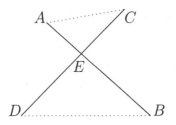

Another well-known problem asks for the proof that given five points in general position, four of them will determine a convex quadrilateral.

The *convex hull* of a set of points is the smallest convex polygon containing all the points. Consider the convex hull of the five given points. If it is a quadrilateral, there is nothing further to prove. If it is a pentagon, omitting any one of the points will yield a desired quadrilateral. The interesting case is when it is a triangle.

Let the points be A, B, C, D and E, with D and E inside triangle ABC. The line through D and E must intersect two sides of the triangle, say AB at P and AC at Q, with D between P and E. Then $BDEC$ is a convex quadrilateral.

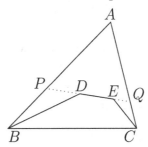

The partition of a polygon into non-overlapping triangles is called a *triangulation*. If a convex n-gon is triangulated by non-intersecting diagonals, then the number of diagonals is $n-3$ and the number of triangles is $n-2$. Note that there is at least one triangle in any such triangulation which is determined by two adjacent sides of the polygon.

For $n = 3$, the only triangle is of the desired type. For $n \geq 4$, the first diagonal drawn divides the polygon into two, and each subpolygon contains one such triangle by induction. Thus we can conclude that for $n \geq 4$, there are at least two triangles in any partition which are determined by two adjacent sides of the polygon.

Combinatorial geometry problems often involve coloring. We present here a famous result.

Sperner's Lemma.
In triangle ABC, A is painted red, B green and C blue. ABC is triangulated, and the vertices of all triangles in the partition are also painted in these three colors. All vertices on BC are painted green or blue, all vertices on CA are painted blue or red and all vertices on AB are painted red or green. Then there is a triangle in the partition whose vertices are painted in different colors.

The diagram below shows an example of a triangulation of ABC. All red vertices are marked A, all green vertices are marked B and all blue vertices are marked C.

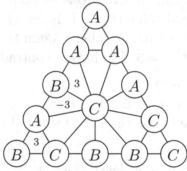

Consider any triangle in the triangulation. An edge of it is assigned the value 0 if the end points have the same color. Otherwise, it is assigned the value ± 1. For an edge AB, we take the plus sign if in going from A to B around the perimeter of the triangle, we are going in the clockwise direction. Otherwise, we take the minus sign.

The sum of values of the three edges is the value assigned to the triangle. Note that a triangle with three vertices in different colors has value ±3. Otherwise, it has value 0. In the diagram above, the triangles with values ±3 are marked in their interiors by these numbers.

To prove Sperner's Lemma, we use an indirect argument. Let S be the sum of all the values of the triangles. Suppose to the contrary that there are no triangles whose vertices are painted in three different colors. Then we have $S = 0$.

The perimeter of the triangulation consists of the three sides BC, CA and AB of the original triangle. Each edge AB on the side AB has value 1 while each edge BA on this side has value -1. Since we start with A and end up with B, the sum of the values of these edges is 1. It follows that the sum of the values of all edges on the perimeter is 3. Since each edge on the perimeter belongs to one triangle, the net contribution of the edges on the perimeter to S is 3.

Each edge in the interior of the triangulation belongs to two triangles. The net contribution to S of each edge is 0. This is obvious if its assigned value is 0. If it is ±1, the contribution is 1 to one triangle and -1 to the other. Again the net contribution is 0. It follows that $S = 3$. We have a contradiction.

Problem F1(1965-2).
Among eight points on or inside a circle, prove that there exist two whose distance is less than the radius of the circle.

Discussion:
Among the eight points, at least seven do not coincide with the center O of the circle. They determine seven radii of the circle, some of which may coincide. Apply the Pigeonhole Principle.

Problem F2(1968-2).
Let n be a positive integer. Inside a circle with radius n are $4n$ segments of length 1. Prove that given any straight line, there exists a chord of the circle, either parallel or perpendicular to the given line, that intersects at least two of the segments.

Discussion:
Consider the projections of the segments on the given line and a line perpendicular to it. Apply the Triangle Inequality.

Problem F3(1983-3).
Let n be a positive integer. P_1, P_2, $\ldots$, P_n and Q are points in the plane with no three on the same line. For any two different points P_i and P_j, there exists a third point P_k such that Q lies inside triangle $P_i P_j P_k$. Prove that n is odd.

Discussion:
From Q, construct rays to P_1, P_2, $\ldots$, P_n, and we may assume that they are in clockwise order. For $1 \leq k \leq n$, let QR_k be the ray opposite to QP_k. Prove that between two adjacent rays QP_i and QP_{i+1}, there exists a ray QR_j between them.

Problem F4(1971-2).
In the plane are 22 points no three of which lie on the same straight line. Prove that they can be joined in pairs by 11 segments such that these segments intersect one another at least 5 times.

Discussion:
Use the fact that given 5 points in general position, 4 of them will determine a convex quadrilateral whose diagonals will intersect.

Problem F5(1994-3).
For $1 \leq k \leq n$, the set H_k consists of k pairwise disjoint intervals of the real line. Prove that among the intervals that form the sets H_k, one can find $\lfloor \frac{n+1}{2} \rfloor$ pairwise disjoint ones, each of which belongs to a different set H_k.

Discussion:
A natural approach is to use induction on n. In reducing $n + 1$ sets to n sets, we have to eliminate one set. The Extremal Value Principle may help us choose which set to eliminate. We must still modify the remaining sets so that they contain the right numbers of intervals. We can then apply the induction hypothesis.

Problem F6(1989-3).
From an arbitrary point (x, y) in the coordinate plane, one is allowed to move to $(x, y+2x)$, $(x, y-2x)$, $(x+2y, y)$ or $(x, x-2y)$. However, one cannot reverse the immediately preceding move. Prove that starting from the point $(1, \sqrt{2})$, it is not possible to return there after any number of moves.

Discussion:
Call a point (x, y) irrational if $\frac{y}{x}$ is irrational. Starting from $(1, \sqrt{2})$ or any other irrational point for that matter, we can only move to other irrational points. If we never revisit any point, the desired conclusion follows. Henceforth we assume that there is a cycle joining distinct irrational points (x_0, y_0), (x_1, y_1), $\ldots$, (x_n, y_n) back to (x_0, y_0). Try to find a contradiction.

Problem F7(1985-1).
Let n be a positive integer. The convex $(n+1)$-gon $P_0 P_1 \ldots P_n$ is divided by non-intersecting diagonals into $n-1$ triangles. Prove that these triangles can be numbered from 1 to $n-1$ such that P_i is a vertex of the triangle numbered i for $1 \le i \le n - 1$.

Discussion:
This can be established by a simple inductive argument.

Problem F8(1967-2).
A convex polygon is divided into triangles by non-intersecting diagonals. If each vertex of the polygon is a vertex of an odd number of these triangles, prove that the number of vertices of the polygon is divisible by 3.

Discussion:
Use mathematical induction on n and suppose the result holds for $n = 3, 4, \ldots, k$ for some $k \geq 4$. Consider a $(k+1)$-gon. If the hypothesis is false, there is nothing to prove. Hence assume that the hypothesis is true. Define a partial polygon as a polygon obtained by dividing the $(k + 1)$-gon with a diagonal. Since $k + 1 \geq 5$, there is at least one partial polygon with more than 3 sides. Prove that a partial polygon which is not a triangle has at least 5 sides.

Problem F9(1994-2).
Consider the diagonals of a convex n-gon.

(a) Prove that if any $n - 3$ of them are omitted, there are $n - 3$ remaining diagonals that do not intersect inside the polygon.

(b) Prove that one can always omit $n - 2$ diagonals such that among any $n - 3$ of the remaining diagonals, there are two which intersect inside the polygon.

Discussion:

(a) It is useful to consider the more general result that if any k diagonals, $k \leq n-3$, are omitted, there are $n-3$ remaining diagonals that do not intersect inside the polygon. Use induction on n.

(b) From the convex n-gon $A_1 A_2 \ldots A_n$, we omit the $n - 2$ diagonals $A_n A_2$, $A_n A_3$, $\ldots$, $A_n A_{n-2}$ and $A_{n-1} A_1$. Consider the resulting convex $(n - 1)$-gon $A_1 A_2 \ldots A_{n-1}$ with none of its diagonals omitted.

Problem F10(1996-3).
For integers $n \geq 3$ and $k \geq 0$, mark some of the diagonals of a convex n-gon. We wish to choose a polygonal line consisting of $2k + 1$ marked diagonals and not intersecting itself.

(a) Prove that this is always possible if $2kn + 1$ diagonals are marked.

(b) Prove that this may not be possible if kn diagonals are marked.

Discussion:
We define the length of a diagonal as ℓ if the shorter way from one endpoint of the diagonal along the perimeter of the n-gon to the other endpoint passes over ℓ sides of the n-gon.

(a) Use mathematical induction on k to prove that if any $2kn + 1$ diagonals of a convex n-gon are marked, we can choose an *open* polygonal line which consists of $2k + 1$ of the marked diagonals and does not intersecting itself.

(b) Let $A_1 A_2 \ldots A_r$ be the longest polygonal line not intersecting itself that can be chosen from the marked diagonals. Then the entire polygonal line lies on one side of $A_1 A_2$, and also on one side of $A_{r-1} A_r$. Mark additional diagonals according to whether n is odd or even.

Problem F11(1978-2).
Let n be a positive integer. The vertices of a convex n-gon are painted so that adjacent vertices have different colors. Prove that if n is odd, then the polygon can be divided into triangles by non-intersecting diagonals such that none of these diagonals has its endpoints painted in the same color.

Discussion:
Since n is odd, $n = 2k - 1$ for some $k \geq 2$. Use mathematical induction on k. In the inductive step, consider a convex $(2k + 1)$-gon $A_1 A_2 \ldots A_{2k+1}$ with vertices painted so that adjacent vertices have different colors. Prove that there exists a diagonal which skips over exactly one vertex which joins two vertices of different colors. Try to cut off some triangles so that the induction hypothesis can be applied.

Problem F12(1970-3).
On the plane are a number of points no three of which lie on the same straight line. Every two of these points are joined by a segment. Some segments are painted red, some others are painted blue, while the remaining ones are unpainted. Every two of the points is connected by a unique polygonal path consisting only of painted segments. Prove that each unpainted segment can be painted red or blue so that in any triangle determined by these points, the number of red sides is odd.

Discussion:
Consider the points and the segments as vertices and edges of a complete graph. The edges which have been painted so far is a connected subgraph with no cycles, covering all vertices. Thus it is a spanning tree. When an unpainted edge is painted, a new cycle is completed. Decide how its color may be chosen.

Problem F13(1992-3).
Given a finite number of points in the plane, no three of which are collinear, prove that they can be painted in two colors so that there is no half-plane that contains exactly three given points of one color and no points of the other color.

Discussion:
Consider the convex hull H of all the points. Paint those on the boundary of H red and those in the interior of H black. A half-plane which contains any black points must contain at least one red point. Suppose some half-plane contains exactly three red points. Let them be A_1, B_1 and C_1 that order along the boundary of H. Then none of the points is inside triangle $A_1B_1C_1$. Repaint B_1 black. Continue to modify the coloring scheme until the desired result is obtained.

Problem F14(1991-3).
Given are 998 red points in the plane, no three on a line. A set of blue points is chosen so that every triangle with all three vertices among the red points contains a blue point in its interior. What is the minimum size of a set of blue points which works regardless of the positions of the red points?

Discussion:
Put 995 red points inside the triangle determined by the other 3 and use them as vertices in a triangulation of the overall triangle into 1991 triangles. Then prove that 995 points are sufficient.

Problem F15(1972-3).
There are four houses in a square plot which is 10 kilometers by 10 kilometers. Roads parallel to the sides of the plot are built inside the plot so that from each house, it is possible to travel by roads to both the north edge and the south edge of the plot. Prove that the minimum total length of the roads is 25 kilometers.

Discussion:
Let the square plot be bounded by $x = 0$, $x = 10$, $y = 0$ and $y = 10$. Consider a special case where the houses are at $(0,2.5)$, $(2.5,10)$, $(7.5,0)$ and $(10,7.5)$. Find the shortest network which works for them, and prove that this result is optimal.

Problem F16(1970-1).

Let n be a positive integer. An n-gon on the plane is not necessarily convex, but non-adjacent sides do not intersect. Determine the maximum number of acute angles of this n-gon.

Discussion:
The diagram below shows that for $n = 3$, 4, 5, 6, 7 and 8, the
n-gon can have as many as 3, 3, 4, 5, 5 and 6 acute angles. Gen-
eralize the construction and prove that these values are optimal.

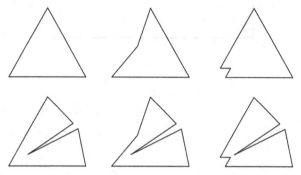

Problem F17(1974-2).

The lengths of the sides of an infinite sequence of squares are 1,
$\frac{1}{2}$, $\frac{1}{3}$, and so on. Determine the length of the side of the smallest
square which can contain all squares in the sequence.

Discussion:
Clearly, we should pack the squares in descending order of size,
starting with the 1×1 and $\frac{1}{2} \times \frac{1}{2}$ squares. It would appear that
the side length of the smallest square box must be at least $1\frac{1}{2}$.
This is obvious if the sides of those two squares are parallel to
the sides of the box, but this is not necessarily the case. Try to
show that even then, the side length of the box is still at least
$1\frac{1}{2}$. Now find an algorithm which can pack all the squares in
such a box. The first seven squares may be placed as shown in
the diagram below.

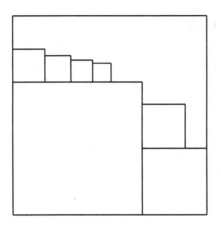

Part III: Solutions

1964

Problem 1.

$PABC$ is a triangular pyramid with $AB = BC = CA$ and $PA = PB = PC$. Another triangular pyramid congruent to $PABC$ is glued to it along the common base ABC to obtain a hexahedron with five vertices such that the dihedral angle between any two adjacent faces is the same. Determine the ratio $PQ : BC$ where Q is the fifth vertex.

Solution:
We first assume that such a· hexahedron exists. Let O be the center of ABC. Then $BC : AO = 3 : \sqrt{3}$. Let F be the midpoint of AB. Then AB is perpendicular to both PF and QF, so that $\angle PFQ$ is the dihedral angle between the faces PAB and QAB. Let R be the foot of perpendicular from B to AP. By symmetry, R is also the foot of perpendicular from C to AP, so that $\angle BRC$ is the dihedral angle between the faces PAB and PCB. By hypothesis, $\angle PFQ = \angle BRC$, so that the isosceles triangles APQ and RBC are similar. It follows that $\frac{PQ}{BC} = \frac{PF}{BR}$. The area of triangle PAB is given by $\frac{1}{2}AB \cdot PF = \frac{1}{2}AP \cdot BR$. Hence $\frac{AP}{AB} = \frac{PF}{BR} = \frac{PQ}{BC}$. Since $AB = BC$, $AP = PQ$ so that triangle APQ is equilateral. Note that O is the midpoint of PQ. Hence $AO : PQ = \sqrt{3} : 2$, so that $PQ : BC = 2 : 3$. We can now justify the existence of such a hexahedron by constructing it. Take an equilateral triangle APQ, rotate it $120°$ clockwise about PQ and let B be the image of A. Rotate APQ $120°$ counterclockwise about PQ and let C be the image of A. Then the hexahedron $P - ABC - Q$ has the desired property.

Problem 2.
At a party, every boy dances with at least one girl, but no girl dances with every boy. Prove that there exist two boys and two girls such that each of these two boy has danced with exactly one of these two girls.

Solution:
Consider some boy Ace. He dances with some girl Bea, and she does not dance with some boy Cec. Among the girls who have danced with Cec, we seek Dee who has not danced with Ace. This can be accomplished by choosing Ace to be a boy who has danced with the lowest number of girls. If Dee does not exist, then Ace has danced with every girl who has danced with Cec, in addition to Bea who has not danced with Cec either. This contradicts the minimality assumption on Ace. Hence Dee must exist.

Problem 3.
Prove that for any positive real numbers a, b, c and d,

$$\sqrt{\frac{a^2 + b^2 + c^2 + d^2}{4}} \geq \sqrt[3]{\frac{abc + bcd + cda + dab}{4}}.$$

Solution:
We first prove that $\sqrt{\dfrac{a^2 + b^2 + c^2 + d^2}{4}} \geq \dfrac{a + b + c + d}{4}$. This follows from

$$\frac{a^2 + b^2 + c^2 + d^2}{4} - \left(\frac{a + b + c + d}{4}\right)^2$$

$$= \frac{1}{16}(3(a^2 + b^2 + c^2 + d^2) - 2(ab + ac + ad + bc + bd + cd))$$

$$= \frac{1}{16}((a - b)^2 + (a - c)^2 + (a - d)^2$$
$$+ (b - c)^2 + (b - d)^2 + (c - d)^2))$$

$$\geq 0.$$

By the Arithmetic-Geometric Means Inequality, we have

$$\frac{abc + bcd + cda + dab}{4}$$

$$= \frac{1}{2}\left(ab\frac{c+d}{2} + cd\frac{a+b}{2}\right)$$

$$\leq \frac{1}{2}\left(\left(\frac{a+b}{2}\right)^2\frac{c+d}{2} + \left(\frac{c+d}{2}\right)^2\frac{a+b}{2}\right)$$

$$= \left(\frac{a+b}{2}\right)\left(\frac{c+d}{2}\right)\frac{a+b+c+d}{4}$$

$$\leq \left(\frac{a+b+c+d}{4}\right)^2\frac{a+b+c+d}{4}$$

$$= \left(\frac{a+b+c+d}{4}\right)^3$$

$$\leq \left(\sqrt{\frac{a^2+b^2+c^2+d^2}{4}}\right)^3 .$$

This is equivalent to the desired inequality.

1965

Problem 1.
Determine all integers a, b and c such that

$$a^2 + b^2 + c^2 + 3 < ab + 3b + 2c.$$

Solution:
Since both sides are integers, we have $a^2+b^2+c^2+4 \leq ab+3b+2c$. This may be rewritten as

$$0 \geq \left(a - \frac{b}{2}\right)^2 + 3\left(\frac{b}{2} - 1\right)^2 + (c - 1)^2.$$

It follows that $a = \frac{b}{2} = 1$ and $c = 1$, so that $(a, b, c) = (1, 2, 1)$ is the only possibility.

Problem 2.
Among eight points on or inside a circle, prove that there exist two whose distance is less than the radius of the circle.

Solution:
Among the eight points, at least seven do not coincide with the center O of the circle. They determine seven radii of the circle, some of which may coincide. Since the sum of the angles between consecutive radii is $360°$, at least one angle is less than $60°$. Let the radii forming this angle be determined by the points P and Q. Then $\angle POQ$ is strictly less than one other angle of triangle OPQ. Hence PQ is less than OP or OQ, either of which is at most the radius of the circle. The desired conclusion follows.

Problem 3.

The base $ABCD$ and the top $EFGH$ of a hexahedron are both squares. The lateral edges AE, BF, CG and DH have equal length. The circumradius of $EFGH$ is less than the circumradius of $ABCD$, which is in turn less than the circumradius of $ABFE$. Prove that the shortest path on the surface of this hexahedron going from A to G passes through only the lateral faces.

Solution:

We first derive some properties of a lateral face, say $ABFE$. Let P be the foot of perpendicular from E to the base $ABCD$, and Q be the foot of perpendicular from E to AB. Then APQ is a right isosceles triangle. We have $EQ > PQ + AQ$ so that $\angle EAQ > \angle AEQ$. Hence $\angle EAB > 45°$ so that $\angle BFG < 135°$. The circumradius R of $ABEF$ is greater than $\dfrac{AB}{\sqrt{2}}$. We have

$AB = 2R\sin AFB > \sqrt{2}AB\sin AFB$ so that $\dfrac{1}{\sqrt{2}} > \sin AFB$.

Hence $\angle AFB < 45°$.

By symmetry, there are three possible types of paths from A to G, crossing CD, EF and BF respectively. In each case, we unfold the surface of the hexahedron along the specified edge and connect A and G by a line segment.

In the first case, AG clearly intersects CD.

In the second case, if AG misses EF, then it intersects EH so that $45° > \angle HGA = \angle GAB > \angle EAB$, which is a contradiction.

In the third case, AG will intersect BF since

$$\angle AFG = \angle AFB + \angle BFG < 45° + 135° = 180°.$$

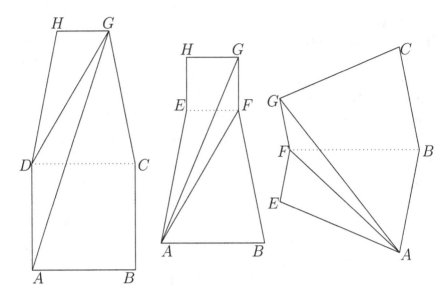

We now compare the lengths of the segment AG in the three cases. Note that $\angle ADG$ in the first case is equal to $\angle AFG$ in the second case. Since $DG = FA$ and $AD > FG$, AG is longer in the first case than in the second. Compare triangle FAG in the last two cases. Since AF and FG are fixed, we can conclude that AG is shorter in the third case than in the second case if we can show that $\angle AFG$ is smaller in the third case than in the second case. In the second case, this is equal to $90° + \angle EFA$. In the third case, we have $2\angle BFE - \angle EFA$. The desired result follows since

$$45° + \angle EFA > \angle AFB + \angle EFA = \angle FBE.$$

1966

Problem 1.
Do there exist five points A, B, C, D and E in space such that

$$AB = BC = CD = DE = EA \quad \text{and}$$

$$\angle ABC = \angle BCD = \angle CDE = \angle DEA = \angle EAB = 90°?$$

Solution:
Suppose such a pentagon $ABCDE$ exists. We may take the length of each side to be 2. Then the length of each diagonal is $2\sqrt{2}$. Place the right isosceles triangle BCD on a horizontal plane and complete the square $BCDF$. Then A lies on the plane through B perpendicular to BC. Since $AC = AD$, A also lies on the plane through the midpoint of CD and perpendicular to CD. The intersection of these two planes is a vertical line through the midpoint G of BF. Since $AB = 2$ and $BG = 1$, we have $AG = \sqrt{3}$. Since $CG = DG = \sqrt{5}$, we indeed have $AC = AD = 2\sqrt{2}$. Similarly, E lies on the vertical line through the midpoint H of DF, with $EH = \sqrt{3}$. If A and E are on opposite sides of the plane BCD, $AE = \sqrt{14} > 2$ and, if they are on the same side, $AE = \sqrt{2} < 2$. Both cases contradict $AE = 2$.

Problem 2.
Let n be any positive integer. Prove that the first n digits after the decimal point of the decimal expansion of the real number $(5 + \sqrt{26})^n$ are identical.

Solution:
When $(5+\sqrt{26})^n + (5-\sqrt{26})^2$ is expanded, the irrational terms cancel out, so that the expression is equal to an integer. Note that $5 < \sqrt{26} < 5.1$ since $25 < 26 < 26.01$. It follows that $|5-\sqrt{26}| < \frac{1}{10}$. Suppose n is odd. Then $(5+\sqrt{26})^n$ is $|(5-\sqrt{26})^n|$ more than an integer. Since $|(5 - \sqrt{26})^n| < (\frac{1}{10})^n$, the first n digits after the decimal point in the decimal expansion of $(5 + \sqrt{26})^n$ are all 0s. Similarly, if n is even, the first n digits after the decimal point are all 9s.

Problem 3.
Do there exist two infinite sets of non-negative integers such that every non-negative integer is expressible as the sum of one element from each set in a unique way?

Solution:
We construct two sets with the desired properties. One set consist of all non-negative integers in which all odd-placed digits are 0s, such as 0, 10, 2010, and so on. The other set consists of all non-negative integers in which all even-placed digits are 0s, such as 0, 102, 10201, and so on. Given any non-negative integers, make two copies of it. In the first copy, replace all odd-placed digits with 0s, and in the second copy, replace all even-placed digits with 0s. The number is clearly the sum of the two modified copies, which belong respectively to the two sets. Since there are no carry over in the addition of two such numbers, the expression is unique.

1967

Problem 1.

In a set of integers which contains both positive and negative elements, the sum of any two elements, not necessarily distinct, is also in the set. Prove that the difference between any two elements is also in the set.

Solution:

We first prove that for any integer c in the set, nc is in the set for any positive integer n. We use induction on n. For $n = 1$, $nc = c$ is in the set. Suppose nc is in the set for some $n \geq 1$. Then $(n + 1)c = nc + c$ is also in the set. This completes the inductive argument. Since the elements are all integers, the set contains a least positive integer a and a largest negative integer b. Now $b < a + b < a$. If $a + b$ is positive, it contradicts the minimality assumption on a. If it is negative, it contradicts the maximality assumption on b. Hence we must have $a + b = 0$ or $b = -a$. It follows that the set contains all integral multiples of a, positive or otherwise. We claim that the set consists only of integral multiples of a, in which case it contains the difference of every two elements in the set. Suppose to the contrary that the set contains a non-multiple of a. Then it has the form $qa + r$ where r is an integer such that $0 < r < a$. Now $(-q)a$ is in the set, as is $r = (qa + r) + (-q)a$. However, this contradicts the minimality assumption on a, and the claim is justified.

Problem 2.

A convex polygon is divided into triangles by non-intersecting diagonals. If each vertex of the polygon is a vertex of an odd number of these triangles, prove that the number of vertices of the polygon is divisible by 3.

Solution:

We use mathematical induction on n. For $n = 3$, each vertex is a vertex of exactly 1 triangle, and 3 is divisible by 3. For $n = 4$, two of the vertices are vertices of exactly 2 triangles, so that the hypothesis is false. Suppose the result holds for $n = 3, 4, \ldots, k$ for some $k \geq 4$. Consider a $(k + 1)$-gon. If the hypothesis is false, there is nothing to prove. Hence we assume that the hypothesis is true. We define a partial polygon as a polygon obtained by dividing the $(k + 1)$-gon with a diagonal. Each diagonal corresponds to two partial polygons. A partial polygon may have 3 sides. Since $k + 1 \geq 5$, there is at least one partial polygon with more than 3 sides. A partial polygon cannot have exactly 4 sides, as otherwise one of its two vertices not on the dividing diagonal will be a vertex of exactly 2 triangles in the overall triangulation. Let m be the minimum number of sides of a partial polygon which is not a triangle. Then $m \geq 5$. Let this polygon be $A_1 A_2 \ldots A_m$. Then $A_1 A_i A_m$ is a triangle for some i, $2 \leq i \leq m - 1$. In the overall triangulation, both $A_1 A_i$ and $A_i A_m$ cut off partial polygons with less than m sides. Hence both are triangles, so that $i = 3$ and $m = 5$. Consider the $(k - 2)$-gon $A_5 A_6 \ldots A_{k+1} A_1$. The vertices A_6, A_7, $\ldots$, A_{k+1} all satisfy the hypothesis. Since A_1 is a vertex of $A_1 A_2 A_3$ and $A_1 A_3 A_5$, it also satisfies the hypothesis. Similarly, so does A_5. By the induction hypothesis, $k - 2$ is a multiple of 3. It follows that $k + 1$ is also a multiple of 3.

Problem 3.

The sum of the distances from each vertex of a convex quadrilateral to the two sides not containing it is constant. Prove that the quadrilateral is a parallelogram.

Solution:

Denote by (P, θ) the sum of the distances from a point P inside an angle θ to the arms of θ. Let P be a point inside an angle with vertex O and let the line through P perpendicular to the bisector of the angle intersect the arms of the angle at A and B. We first prove an auxiliary result, that as $OA = OB = \ell$ increases, so does $(P, \angle AOB)$. Let d be the distance from A to OB. Then the area of triangle AOB is $\frac{1}{2}\ell d$. On the other hand, it is the sum of the areas of triangles POA and POB, which is $\frac{1}{2}\ell(P, \angle AOB)$. When ℓ increases, d also increases, and so does $(P, \angle AOB)$. It follows that if Q is another point inside $\angle AOB$, then

$$(P, \angle AOB) = (Q, \angle AOB)$$

if and only if Q lies on AB.

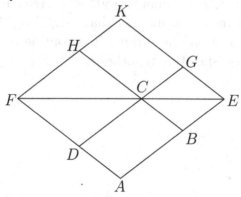

Let $ABCD$ be a convex quadrilateral. Complete the rhombus $AEKF$ with B on AE, D on AF and C on EF. Let G be the point of intersection of CD and KE, and H be the point of intersection of BC and KF.

Note that triangles CBE and CHF are similar, as are triangles CGE and CDF. Hence the quadrilaterals $CBEG$ and $CHFD$ are similar, so that BG is parallel to DH. By hypothesis,

$$(B, \angle ADC) = (C, \angle BAD).$$

Now

$$(C, \angle BAD) = (E, \angle BAD)$$

by our auxiliary result. Also,

$$(E, \angle BAD) = (G, \angle ADC)$$

since both are equal to the distance between EK and AF. Hence

$$(B, \angle ADC) = (G, \angle ADC),$$

so that BG is perpendicular to the bisector of $\angle ADC$. Similarly, DH is perpendicular to the bisector of $\angle ABC$. Since BG is parallel to DH, the two angle bisectors are parallel. Hence each of $\angle BAD$ and $\angle BCD$ is equal to $180° - \frac{1}{2}(\angle ABC + \angle ADC)$. We can prove in an analogous manner that $\angle ABC = \angle ADC$, so that $ABCD$ is indeed a parallelogram. It is routine to verify that a parallelogram satisfies the hypothesis.

1968

Problem 1.

Prove that in any infinite sequence of distinct positive integers, it is not possible for every block of three consecutive terms a, b and c to satisfy $b = \dfrac{2ac}{a+c}$.

Solution:

Note that $\frac{1}{b} = \frac{1}{2}(\frac{1}{a} + \frac{1}{c})$. Thus the existence of such a sequence implies the existence of an infinite arithmetic progression consisting of reciprocals of distinct positive integers. However, the terms of this progression all lie between 0 and 1. This is a contradiction.

Problem 2.

Let n be a positive integer. Inside a circle with radius n are $4n$ segments of length 1. Prove that given any straight line, there exists a chord of the circle, either parallel or perpendicular to the given line, that intersects at least two of the segments.

Solution:

Let ℓ be the given straight line and let m be a line perpendicular to ℓ. For $1 \leq i \leq 4n$, let the lengths of the projections of the ith segment on ℓ and m be a_i and b_i respectively. By the Triangle Inequality, we have $a_i + b_i \geq 1$. The projections of the circle on ℓ and m both have length $2n$. Since

$$(a_1 + b_1) + (a_2 + b_2) + \cdots + (a_{4n} + b_{4n}) \geq 4n,$$

the projections of at least two segments on ℓ or on m must have a common point. We may assume that this point is on ℓ. Then the line through this point perpendicular to ℓ will intersect two of the segments.

Problem 3.

Let $n > k > 0$ be integers. For each arrangement of n white balls and n black balls in a row, count the number of pairs of adjacent balls of different colors. Prove that the number of arrangements for which the count is $n - k$ is equal to the number of arrangements for which the count is $n + k$.

Solution:

We first consider the case where $n - k$ is odd. Then $n + k$ is also odd. Let $n - k = 2a + 1$ and $n + k = 2b + 1$ for some positive integers a and b. Then $a + b = n - 1$. Divide any arrangement of the $2n$ balls with $2a + 1$ changes of color into $2a + 2$ monochromatic segments. Divide the $2n$ balls into two monochromatic arrangements, each with n balls, by combining the $a+1$ white segments and the $a+1$ black segments separately. Conversely, we can divide an arrangement of n white balls into $a + 1$ non-empty segments with a dividers. This can be done in $\binom{n-1}{a}$ ways. Similarly, we can divide an arrangement of n black balls into $a + 1$ non-empty segments in $\binom{n-1}{a}$ ways. We now merge them into an arrangement with $2a + 2$ segments of alternating colors. Since the merged arrangement can begin with either a white segment or a black segment, the number of such arrangements is $2\binom{n-1}{a}^2$. Similarly, the number of arrangements of $2n$ balls with $2b+1$ changes of color is $2\binom{n-1}{b}^2$. Note that since $a+b = n-1$, $\binom{n-1}{a} = \binom{n-1}{b}$, and the desired conclusion follows. The case where $n - k$ is even can be handled in a similar manner. Let $n - k = 2a$ and $n + k = 2b$ for some positive integers a and b. Then $a + b = n$. The number of arrangements with $2a$ changes of color is $2\binom{n-1}{a}\binom{n-1}{a-1}$ while the number of arrangements with $2b$ changes of color is $2\binom{n-1}{b}\binom{n-1}{b-1}$. Now $\binom{n-1}{a} = \binom{n-1}{b-1}$ and $\binom{n-1}{a-1} = \binom{n-1}{b}$, and the desired conclusion follows.

1969

Problem 1.

Let n be a positive integer. Prove that if $2 + 2\sqrt{28n^2 + 1}$ is an integer, then it is the square of an integer.

Solution:

Let $m = 1 + \sqrt{28n^2 + 1}$. Then $m^2 - 2m = 28n^2$. Hence $m = 2k$ for some positive integer k, so that $7n^2 = k^2 - k = k(k - 1)$. Since k and $k - 1$ are relatively prime, one of them is a square and the other is 7 times a square. Suppose $k - 1 = h^2$ for some positive integer h. Then k is divisible by 7, and we have $h^2 \equiv 6$ (mod 7). However, a square can only be congruent modulo 7 to 0, 1, 2 or 4. Hence we must have $k = h^2$ instead. Now $2 + 2\sqrt{28n^2 + 1} = 2m = 4k = (2h)^2$ as desired.

Problem 2.

Let the lengths of the sides of a triangle be a, b and c, and the measures of the opposite angles be α, β and γ respectively. Prove that the triangle is equilateral if

$$a(1 - 2\cos\alpha) + b(1 - 2\cos\beta) + c(1 - 2\cos\gamma) = 0.$$

Solution:

The given condition is equivalent to

$$\sin\alpha(1 - 2\cos\alpha) + \sin\beta(1 - 2\cos\beta) + \sin\gamma(1 - 2\cos\gamma) = 0,$$

so that $\sin\alpha + \sin\beta + \sin\gamma = \sin 2\alpha + \sin 2\beta + \sin 2\gamma$. Since $\alpha + \beta + \gamma = 180°$, we have

$$
\begin{aligned}
&\sin\alpha + \sin\beta + \sin\gamma \\
&= 2\sin\frac{\alpha + \beta}{2}\cos\frac{\alpha - \beta}{2} + 2\sin\frac{\alpha + \beta}{2}\cos\frac{\alpha + \beta}{2} \\
&= 2\sin\frac{\alpha + \beta}{2}\left(\cos\frac{\alpha - \beta}{2} + \cos\frac{\alpha + \beta}{2}\right) \\
&= 4\cos\frac{\alpha}{2}\cos\frac{\beta}{2}\cos\frac{\gamma}{2}.
\end{aligned}
$$

On the other hand,

$$
\begin{aligned}
&\sin 2\alpha + \sin 2\beta + \sin 2\gamma \\
={}& 2\sin(\alpha + \beta)\cos(\alpha - \beta) + \sin 2\gamma \\
={}& 2\sin\gamma\cos(\alpha - \beta) + 2\sin\gamma\cos\gamma \\
={}& 2\sin\gamma(\cos(\alpha - \beta) - \cos(\alpha + \beta)) \\
={}& 4\sin\alpha\sin\beta\sin\gamma.
\end{aligned}
$$

We now have

$$
\begin{aligned}
\cos\frac{\alpha}{2}\cos\frac{\beta}{2}\cos\frac{\gamma}{2} &= \sin\alpha\sin\beta\sin\gamma \\
&= \frac{1}{8}\sin\frac{\alpha}{2}\sin\frac{\beta}{2}\sin\frac{\gamma}{2}\cos\frac{\alpha}{2}\cos\frac{\beta}{2}\cos\frac{\gamma}{2}.
\end{aligned}
$$

Since $\cos\frac{\alpha}{2}\cos\frac{\beta}{2}\cos\frac{\gamma}{2} \neq 0$,

$$
\begin{aligned}
\frac{1}{8} &= \sin\frac{\alpha}{2}\sin\frac{\beta}{2}\sin\frac{\gamma}{2} \\
&= \frac{1}{2}\left(\cos\frac{\alpha - \beta}{2} - \cos\frac{\alpha + \beta}{2}\right)\sin\frac{\gamma}{2} \\
&= \frac{1}{2}\left(\cos\frac{\alpha - \beta}{2} - \sin\frac{\gamma}{2}\right)\sin\frac{\gamma}{2}.
\end{aligned}
$$

Hence $\sin\frac{\gamma}{2}$ is a real root of $x^2 - \cos\frac{\alpha-\beta}{2}x + \frac{1}{4} = 0$. The discriminant of this quadratic equation is $\cos^2\frac{\alpha-\beta}{2} - 1 \leq 0$. Since there is a real root, we must have $\cos^2\frac{\alpha-\beta}{2} = 1$ so that $\alpha = \beta$. It follows that $\sin\frac{\gamma}{2} = \frac{1}{2}\cos\frac{\alpha-\beta}{2} = \frac{1}{2}$ so that $\gamma = 60°$. Hence $\alpha = \beta = 60°$ also, so that the triangle is indeed equilateral.

Problem 3.

A $1 \times 8 \times 8$ block consists of 64 unit cubes such that exactly one face of each cube is painted black. The initial arrangement is arbitrary. In a move, we may rotate a row or a column of 8 cubes about their common axis. Prove that after a finite number of such moves, we can obtain an arrangement in which all the black faces are on top.

Solution:

Label the rows 1 to 8 from bottom to top and the columns a to h from left to right. A cube is said to face in the direction of its black face. Each cube may be facing front, back, left, right, top or bottom. We first make a1 face front. If it is facing back, top or bottom, rotate row 1 so that a1 faces front. If it is facing left or right, we first rotate column a until a1 faces top or bottom before rotating row 1. Then further rotation of column a leaves a1 facing front. Hence we can make the rest of column a face front. Now we rotate the rows to make column a face top. This is followed by rotating column a to make column a face left. Then further rotations of the rows leave column a facing left. In the same way, we can make each column face left. The columns to the left remain facing left since they are not rotated. When all the cubes are facing left, rotations of the columns can make them all face top.

1970

Problem 1.
Let n be a positive integer. An n-gon on the plane is not necessarily convex, but non-adjacent sides do not intersect. Determine the maximum number of acute angles of this n-gon.

Solution:
Let k be the number of acute angles of such an n-gon. Then the sum of these k angles is less than $90k°$. The sum of the remaining $n - k$ angles is less than $360(n - k)°$. Now the sum of all n angles is $180(n - 2)° > 90k° + 360(n - k)°$. Hence $2(n - 2) > k + 4(n - k)$ which simplifies to $3k < 2n + 4$. We consider three cases.

Case 1. $n = 3m$ for some positive integer m.
Then $k \leq 2m + 1$. Take a sector OC_0C_m of a circle with a $60°$ at the center O, and divide the arc C_0C_m into m equal parts by the points C_1, C_2, ..., C_{m-1}. Then we have a convex $(m + 2)$-gon $OC_0C_1 \ldots C_m$ with three acute angles, namely, $\angle OC_0C_1$, $\angle C_{m-1}C_mO$ and $\angle C_mOC_0$. Take a point D_1 on OC_1. Since $\angle D_1C_1C_0 < 90°$ and $\angle D_1C_1C_2 < 90°$, we can choose E_1 on C_0C_1 and F_1 on C_1C_2 close enough to C_1 so that $\angle D_1E_1C_0 < 90°$ and $\angle D_1F_1C_2 < 90°$. Removing the kite $D_1E_1C_1F_1$ yields a non-convex $(m + 4)$-gon $OC_0E_1D_1F_1C_2 \ldots C_m$ with two additional acute angles, namely $\angle D_1E_1C_0$ and $\angle D_1F_1C_2$. For $2 \leq i \leq m-1$, we can remove kites $D_iE_iC_iF_i$ similarly situated as and congruent to $D_1E_1C_1F_1$. The resulting non-convex polygon has $m + 2 + 2(m - 1) = 3m$ sides and $3 + 2(m - 1) = 2m + 1$ acute angles. Thus the maximum value is attained. The diagram illustrates the case $m = 3$.

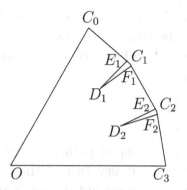

Case 2. $n = 3m + 1$ for some positive integer m.
Then $k \leq 2m+1$. We start with the $3m$-gon constructed in Case 1. Let P be the midpoint of OC_0. Displace it slightly along the perpendicular bisector of OC_0 so that P is inside the original polygon. We now have a $(3m + 1)$-gon still with $2m + 1$ acute angles since $\angle POC_m < \angle C_0OC_m$ and $\angle PC_0E_1 < \angle OC_0E_1$.

Case 3. $n = 3m + 2$ for some positive integer m.
Then $k \leq 2m+2$. We start with the $3m$-gon constructed in Case 1. Let P be a point on OC_0 and Q be a point on OC_m such that $OP = OQ$. Choose their common length small enough so that when we complete the rhombus $OPRQ$, R lies inside the original polygon. Remove this rhombus and we have a $(3m+2)$-gon. We have lost one acute angle, namely, $\angle POQ = 60°$. However, we have gained two new ones, namely, $\angle RPC_0 = \angle RQC_m = 60°$. Once again, the maximum value is attained.

Problem 2.
Determine the probability that five numbers chosen at random from the first 90 positive integers contain two consecutive numbers.

Solution:
The total number of ways of choosing 5 numbers from 90 is $\binom{90}{5}$. Suppose no two of the five numbers $1 \leq a < b < c < d < e \leq 90$ are consecutive. Define $p = b - 1$, $q = r - 2$, $s = d - 3$ and $s = e - 4$. Then we have $1 \leq a < p < q < r < s \leq 86$.

Hence the number of ways of choosing 5 numbers from 90 with no two consecutive is equal to the number of ways of choosing 5 numbers from 86. It follows that the desired probability is

$$1 - \frac{\binom{86}{5}}{\binom{100}{5}} = \frac{106081}{511038}.$$

Problem 3.
On the plane are a number of points no three of which lie on the same straight line. Every two of these points are joined by a segment. Some segments are painted red, some others are painted blue, while the remaining ones are unpainted. Every two of the points is connected by a unique polygonal path consisting only of painted segments. Prove that each unpainted segment can be painted red or blue so that in any triangle determined by these points, the number of red sides is odd.

Solution:
Consider the points and the segments as vertices and edges of a complete graph. The edges which have been painted so far is a connected subgraph with no cycles, covering all vertices. Thus it is a spanning tree. When the first unpainted edge is painted, a new cycle is completed. We choose the color of this first edge so that the number of blue edges on the new cycle is even. Now there are two paths joining any two vertices on the cycle, but the numbers of blue edges on the paths are either both odd or both even. When an unpainted edge is painted subsequently, there are paths joining the endpoints of this edge, and the numbers of blue edges on them have the same parity. Thus we can choose the color for painting so that the number of blue edges on any of these paths is consistently even. Now a triangle is just a cycle of length 3, and the number of blue edges on it is even. Hence it has an odd number of red sides.

1971

Problem 1.

A straight line intersects the sides BC, CA and AB, or their extensions, of triangle ABC at D, E and F respectively. Q and R are the images of E and F under 180° rotations about the midpoints of CA and AB respectively, and P is the point of intersection of BC and QR. Prove that

$$\frac{\sin EDC}{\sin RPB} = \frac{QR}{EF}.$$

Solution:

Construct the altitude AH from A to BC. Let K, L, M and N be the feet of perpendicular from E, F, Q and R to AH respectively. The symmetry between E and Q means that $AM = KH$ and the symmetry between F and R means that $AL = NH$. Hence $MN = AH - AM - NH = H - KH - AL = KL$. Now $\sin EDC = \sin FEK = \dfrac{KL}{EF}$ while $\sin RPB = \sin QRN = \dfrac{MN}{QR}$. It follows that $\dfrac{\sin EDC}{\sin RPB} = \dfrac{QR}{EF}$.

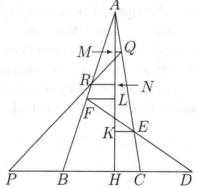

Problem 2.
In the plane are 22 points, no three of which lie on the same straight line. Prove that they can be joined in pairs by 11 segments such that these segments intersect one another at least 5 times.

Solution:
We use the well-known result that given 5 points on the plane, no 3 collinear, 4 of them will determine a convex quadrilateral. Choose a line ℓ_0 with all 22 points on the same side of it, and not parallel to any segment joining two of them. For $1 \le i \le 5$, let ℓ_i be parallel to ℓ_0, such that there are 5 points between ℓ_0 and ℓ_1, 4 points between ℓ_1 and ℓ_2, 4 points between ℓ_2 and ℓ_3, 4 points between ℓ_3 and ℓ_4, and 5 points between ℓ_4 and ℓ_5. By the result mentioned earlier, the 5 points between ℓ_0 and ℓ_1 determine a convex quadrilateral, and its diagonal intersect at a point P_1 between ℓ_0 and ℓ_1. The point not used in the determination of this quadrilateral is added to the 4 points between ℓ_1 and ℓ_2, and they determine a point P_2 of intersection there. We can obtain P_3, P_4 and P_5 in the same way. These are the 5 distinct points we seek since they are separated by ℓ_1, ℓ_2, ℓ_3 and ℓ_4.

Problem 3.
Each of 30 boxes can be opened by a unique key. These keys are then locked at random inside the boxes, with one key in each. Two of the boxes are then broken open simultaneously, and the keys inside may be used to try to open other boxes. Keys retrieved from boxes thus opened may also be used. Determine the probability that all boxes may be opened.

Solution:

More generally, let there be n box-key pairs, numbered 1 to n. We may assume that boxes 1 and 2 are broken. For $1 \le k \le n$, let a_k be the number of the box whose key is in box k. We can decompose the permutation $(a_1, a_2, \ldots, a_n)$ into cycles. Write the numbers in each cycle so that the smallest number within the cycle is in the last place. Arrange the cycles in increasing order of their last numbers. For example, the permutation $\{2, 5, 1, 8, 6, 3, 7, 4\}$ becomes $(2,5,6,3,1)(8,4)(7)$. Now the task is successful if and only if the last number of the last cycle is 1 or 2. Merge the cycle without rearranging the order of the n numbers and obtain the associated permutation of the original permutation. Then the task is successful if and only if the last number of the associated permutations is 1 or 2. For example, merging the cycles in $(2,5,6,3,1)(8,4)(7)$ yields the associated permutation $\{2, 5, 6, 3, 1, 8, 4, 7\}$, and the task fails. There is a one-to-one correspondence between the $n!$ permutations and the $n!$ associated permutations. The probability of the last number of a permutation being 1 or 2 is clearly $\frac{2}{n}$, which is also the probability that the task is successful. For $n = 30$, the probability is $\frac{1}{15}$.

1972

Problem 1.

Prove that $a(b-c)^2 + b(c-a)^2 + c(a-b)^2 + 4abc > a^3 + b^3 + c^3$, where a, b and c are the lengths of the sides of a triangle.

Solution:

We have

$$
\begin{aligned}
& a(b-c)^2 + b(c-a)^2 + c(a-b)^2 + 4abc - (a^3 + b^3 + c^3) \\
={} & a((b-c)^2 - a^2) + b((c-a)^2 - b^2) + c((a+b)^2 - c^2) \\
={} & a(b-c+a)(b-c-a) + b(c-a+b)(c-a-b) \\
& + c(a+b+c)(a+b-c) \\
={} & (a+b-c)(a(b-c-a) - b(c-a+b) + c(a+b+c)) \\
={} & (a+b-c)(c^2 - (a-b)^2) \\
={} & (a+b-c)(b+c-a)(c+a-b) \\
>{} & 0.
\end{aligned}
$$

Problem 2.

In a class with at least 4 students, the number of boys is equal to the number of girls. Consider all arrangements of these students in a row. Let a be the number of arrangements for which it is impossible to divide it into two parts so that the number of boys is equal to the number of girls in each part. Let b be the number of arrangement for which there is a unique partition with this property. Prove that $b = 2a$.

Solution:

We represent the arrangement of the students by a binary sequence, with 0 and 1 representing the opposite genders. We may assume that the first term is 0. A sequence which cannot be partitioned in the desired way is said to be of type A. In such a sequence, the running totals of 0s and 1s are never equal until the end.

A sequence which can be so partitioned in a unique way is said to be of type B. Such a sequence has a unique partition into two subsequences of type A. The sequences of type B form companion pairs, where the first subsequences are identical and in the second subsequences, 0s and 1s are interchanged. It follows that in one sequence of each pair, the second subsequence has 0 as the first term. We now move this 0 in front of the first subsequence. Then the running total of 0s exceeds the running total of 1s up to and including the end of the first subsequence since the first subsequence is of type A. Since the second subsequence is also of type A, the entire sequence is of type A.

Conversely, start with any sequence of type A with at least 4 terms. Then the second term must also be 0. The subsequence which consists only of the first term has one more 0s than 1s. The subsequence which consists of all but the last term also has one more 0s than 1s. Consider the shortest subsequence with at least 2 terms which has one more 0s than 1s. Delete the first 0 and call it the first subsequence, which must be of type A. Restore the deleted 0 to the front of the remaining part of the sequence and call it the second subsequence. This is also of type A since the entire sequence is of type A. It follows that the sequence is also of type B. The one-to-one correspondence between sequences of type A and companion pairs of sequences of type B yields the desired result.

Problem 3.
There are four houses in a square plot which is 10 kilometers by 10 kilometers. Roads parallel to the sides of the plot are built inside the plot so that from each house, it is possible to travel by roads to both the north edge or the south edge of the plot. Prove that the minimum total length of the roads is 25 kilometers.

Solution:

Let the square plot be bounded by $x = 0$, $x = 10$, $y = 0$ and $y = 10$. For $i = 1, 2, 3, 4$, let the ith house be at the point (x_i, y_i), with $x_1 \leq x_2 \leq x_3 \leq x_4$. We first prove that a total length of 25 for the roads is sufficient. Suppose $x_3 - x_2 \leq 5$. Construct a vertical road between x_2 and x_3, or at $x_2 = x_3$. This uses up a length of 10. The horizontal roads from (x_2, y_2) and (x_3, y_3) have total length at most 5, while the horizontal roads from (x_1, y_1) and (x_4, y_4) have total length at most 10. Suppose $x_3 - x_2 > 5$ instead. Construct a vertical road between x_1 and x_2 or at $x_1 = x_2$, and another vertical road between x_3 and x_4 or at $x_3 = x_4$. This uses up a total length of 20. The total length of the four horizontal roads from the houses to the nearer of the two vertical roads is at most $10 - (x_3 - x_2)$, which is at most 5. This establishes sufficiency. We now show that 25 is necessary. Place the houses at $(0,2.5)$, $(2.5,10)$, $(7.5,0)$ and $(10,7.5)$. The network of roads can have at most two components, since the total length of the vertical roads in each component is at least 10. Suppose there are two components. If there are two houses in each component, then the total length of the horizontal roads in each component is at least 2.5, and 25 is indeed necessary. Suppose one component consists of a vertical road through one house, then the total length of the horizontal roads in the other component is at least 7.5, and 25 will not be sufficient. Finally, suppose the network is connected. Then there is a path connecting $(0,2.5)$ and $(2.5,10)$, and another path connecting $(7.5,0)$ and $(10,7.5)$. We consider two cases.

Case 1. These two paths have at most one point in common. Then the total length of the vertical roads in each path is at least 7.5 while the total length of the horizontal roads overall is at least 10. Hence 25 is necessary.

Case 2. These two paths have at least two points in common. To minimize the total length of the roads, the first and the last common points must be connected by a unique subpath common to both paths. Then the path between (0,2.5) and (7.5,0) and the path between (2.5,10) and (10,7.5) have no common points. The same argument in Case 1 shows that 25 is necessary.

1973

Problem 1.

Determine all integers n and k, $n > k > 0$, such that the binomial coefficients $\binom{n}{k-1}$, $\binom{n}{k}$ and $\binom{n}{k+1}$ form an arithmetic progression.

Solution:

Suppose for $1 \leq k \leq n - 1$, we have

$$
\begin{aligned}
0 &= \binom{n}{k-1} - 2\binom{n}{k} + \binom{n}{k+1} \\
&= \frac{n!(k(k+1) - 2(k+1)(n-k+1) + (n-k+1)(n-k))}{(k+1)!(n-k+1)!} \\
&= \frac{n!((n-2k)^2 - n - 2)}{(k+1)!(n-k+1)!}.
\end{aligned}
$$

Then $n = m^2 - 2$ where $m = n - 2k$ or $m = 2k - n$, with $m \geq 2$. In the former case, $k = \frac{n-m}{2} = \binom{m}{2} - 1$. In the latter case, $k = \frac{n+m}{2} = \binom{m+1}{2} - 1$. For $m = 2$, we have $n = 2$ and $k = 0$ or 2, provided that we take $\binom{2}{-1} = \binom{2}{3} = 0$.

Problem 2.

For any point on the plane of a circle other than its center, the line through the point and the center intersects the circle at two points. The distance from the point to the nearer intersection point is defined as the distance from the point to the circle. Prove that for any positive number ϵ, there exists a lattice point whose distance from a given circle with center $(0,0)$ and radius r is less than ϵ, provided that r is sufficiently large.

Solution:
Consider the part of the circle in the first quadrant. Choose a point P on the x-axis such that the vertical line through P intersects the circle at a point C which is not a lattice point, and $OP > r - 1$, where O is the center of the circle. Let B be the first lattice point below C. Then

$$\frac{CP}{OP} = \frac{\sqrt{r^2 - CP^2}}{OP}$$

$$< \frac{\sqrt{2r}}{OP}$$

$$< \frac{\sqrt{2r}}{r - 1}$$

$$< \frac{\sqrt{2r}}{r - \frac{r}{2}}$$

$$= \frac{2\sqrt{2}}{\sqrt{r}}.$$

Let A be the point of intersection of the tangent to the circle at C and the horizontal line passing through B. Then the segment AB intersects the circle, so that the distance of B from the circle is less than AB. Since $BC < 1$, $BD < \frac{AB}{BC}$. Since triangles ABC and OPC are similar, we have $\frac{AB}{BC} = \frac{CP}{OP} < \frac{2\sqrt{2}}{\sqrt{r}}$ provided that $r > 2$. If we also have $r > \frac{8}{\epsilon^2}$, then $AB < \epsilon$ and B is a lattice point whose distance from the circle is less than ϵ.

Problem 3.
Let n be a integer greater than 4. Every three of n planes have a common point, but no four of these planes have a common point. Prove that among the regions into which space is divided by these planes, the number of tetrahedra is not less than $\frac{2n-3}{4}$.

Solution:
We call a point of intersection of three of the n planes a vertex. A plane is said to be special if all the vertices not on it are on the same side of it. We first prove an auxiliary result, that there are at most 3 special planes. Suppose there are 4 special planes. Since every three of them determine a vertex, these 4 planes determine a tetrahedron $ABCD$. Since each is special, all vertices are either on the faces of $ABCD$ or inside it. Note that since $n \geq 5$, there is at least one plane Π other than those which determine $ABCD$. Now Π cannot intersect all six edges of $ABCD$. We may assume that it intersects the line AB at a point E not on the edge AB. Then E is a vertex outside $ABCD$. This is a contradiction. Returning to the main problem, let P be a vertex on one side of a plane Π_0 and closest to Π_0. Let P be determined by Π_1, Π_2 and Π_3. These 4 planes determine a tetrahedron $PQRS$ with Q, R and S on Π_0. Suppose another plane intersects $PQRS$. Then it must intersect at least one of PQ, PR and PS, say PQ at T. Then T is on the same side of Π_0 as P but closer to Π_0, contradicting the minimality assumption on P. Hence $PQRS$ is among the tetrahedra we seek. We say that $PQRS$ is associated with Π_0. Now a special plane has only one tetrahedron associated with it, but all other planes have two. Hence the total number of associated tetrahedra is at least $3 + 2(n - 3) = 2n - 3$. Since each tetrahedron is associated with at most 4 planes, the total number of tetrahedra we seek is at least $\frac{2n-3}{4}$.

1974

Problem 1.
When someone enters a library, she writes down on a blackboard the number of people already in the library at the time. When someone leaves a library, she writes down on a whiteboard the number of people still in the library. Prove that at the end of the day, the numbers on the blackboard are the same as those on the whiteboard, taking into consideration multiplicity but not order.

Solution:
We can represent the visits to the library during the day by a sequence of $+$s and $-$s. The identities of the persons entering or leaving is immaterial. Clearly, the number of $+$s is equal to the number of $-$s. The sequence starts with a $+$ and ends with a $-$. Scan the sequence until the first time we come across a $+$ followed immediately by a $-$. The number on the blackboard generated by this $+$ is clearly equal to the number on the whiteboard generated by this $-$. Remove this $(+, -)$ pair and starting scanning from the beginning of the sequence again. Thus we can remove one pair of $(+, -)$ at a time, until the entire sequence is removed. The desired conclusion follows immediately.

Problem 2.
The lengths of the sides of an infinite sequence of squares are 1, $\frac{1}{2}$, $\frac{1}{3}$, and so on. Determine the length of the side of the smallest square which can contain all squares in the sequence.

Solution:
We first show that a $1\frac{1}{2} \times 1\frac{1}{2}$ square is large enough. Group the squares according to their side lengths as follows: (1), $(\frac{1}{2}, \frac{1}{3})$, $(\frac{1}{4}, \frac{1}{5}, \frac{1}{6}, \frac{1}{7})$, $(\frac{1}{8}, \frac{1}{9}, \ldots, \frac{1}{15})$, $\ldots$, with 2^{k-1} squares in the kth group. The squares in the second group are placed to the right and the squares in the third group are placed on top of the first square, as shown in the diagram.

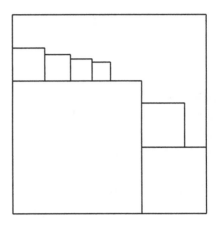

After the squares in the kth group has been placed for some $k \geq 3$, two squares of the $(k+1)$st group are placed on top of each square of the kth group in descending order of size. Note that for $1 \leq i \leq 2^{k-1}$,

$$\frac{1}{2^k + 2i} + \frac{1}{2^k + 2i + 1} < \frac{2}{2^k + 2i} = \frac{1}{2^{k-1} + i},$$

so that the two squares can fit on top of the square below. Finally, since $1 + \frac{1}{4} + \frac{1}{8} + \frac{1}{16} + \cdots < 1 + \frac{1}{2}$, all the squares will fit inside the $1\frac{1}{2} \times 1\frac{1}{2}$ square.

We must still prove that no smaller square will do. We begin with an auxiliary result, that the largest square inside a right triangle shares a vertex with the triangle. Let the triangle be ABC with $\angle C = 90°$. The largest square $KLMN$ inside ABC must have a vertex on each side of it, say K on AC, L on BC and M on AB, as shown in the next diagram. Suppose neither K nor L coincides with C. Let the bisector of $\angle C$ intersect MN at P and KM at its midpoint Q. Since $\angle PCK = 45° = \angle PMK$, $CKPM$ is a cyclic quadrilateral. Now the distance of the center of the circle from KM is shorter than its distance from CP, which means that $CP > KM$. Hence the square with CP as a diagonal is inside ABC and larger than $KLMN$, a contradiction.

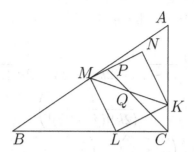

Returning to the main problem, consider the smallest square Z containing a 1×1 square X and a $\frac{1}{2} \times \frac{1}{2}$ square Y. Since they do not overlap, there is a line separating them. If this line is parallel to a side of Z, its side length must be at least $1\frac{1}{2}$. Suppose it is not. Then this line forms two right triangles with the sides of Z, one containing X and the other containing Y. By our auxiliary result and the minimality assumption on Z, the sides of X and Y are both parallel to the sides of Z, and the side length of Z must still be at least $1\frac{1}{2}$.

Problem 3.
Prove that for any real number x and any positive integer k,

$$1 - x + \frac{x^2}{2!} - \frac{x^3}{3!} + \cdots - \frac{x^{2k-1}}{(2k-1)!} + \frac{x^{2k}}{(2k)!} \geq 1.$$

Solution:
Let $P_{2k}(x) = 1 - x + \dfrac{x^2}{2!} - \dfrac{x^3}{3!} + \cdots - \dfrac{x^{2k-1}}{(2k-1)!} + \dfrac{x^{2k}}{(2k)!}$. We will then have $P_{2k}(-x) = 1 + x + \dfrac{x^2}{2!} + \dfrac{x^3}{3!} + \cdots + \dfrac{x^{2k-1}}{(2k-1)!} + \dfrac{x^{2k}}{(2k)!}$, and their product is $P_{2k}(x)P_{2k}(-x) = a_0 + a_1 x + \cdots + a_{4k}x^{4k}$. We have $a_0 = 1$. For $1 \leq j \leq 2k$,

$$
\begin{aligned}
a_j &= \frac{1}{j!} - \frac{1}{(j-1)!} + \frac{1}{2!(j-2)!} - \cdots + (-1)^j \frac{1}{j!} \\
&= \frac{1}{j!}\left(\binom{j}{0} - \binom{j}{1} + \binom{j}{2} - \cdots + (-1)^j \binom{j}{j}\right) \\
&= 0.
\end{aligned}
$$

For $2k + 1 \leq j \leq 4k$, we have

$$
\begin{aligned}
a_j &= (-1)^j \left(\frac{1}{(j - 2k)!(2k)!} - \frac{1}{(j - 2k + 1)!(2k - 1)!} \right. \\
&\quad \left. + \cdots + (-1)^j \frac{1}{(2k)!(j - 2k)!} \right) \\
&= (-1)^j \left(\binom{j}{j - 2k} - \binom{j}{j - 2k + 1} \right. \\
&\quad \left. + \cdots + (-1)^j \binom{j}{2k} \right).
\end{aligned}
$$

If j is odd, then $a_j = 0$. If j is even, then

$$
\begin{aligned}
a_j &= \left(\binom{j - 1}{j - 2k - 1} + \binom{j - 1}{j - 2k} \right) \\
&\quad - \left(\binom{j - 1}{j - 2k} + \binom{j - 1}{j - 2k + 1} \right) \\
&\quad + \cdots + \left(\binom{j - 1}{2k} + \binom{j - 1}{2k + 1} \right) \\
&= \binom{j - 1}{j - 2k} + \binom{j - 1}{2k + 1} \\
&> 0.
\end{aligned}
$$

Hence $P_{2k}(x)P_{2k}(-x) > 0$ for all x. Since $P_{2k}(-x) > 0$ for all $x > 0$, this implies that $P_{2k}(x) > 0$ for all $x > 0$. Clearly, $P_{2k}(x) > 0$ for all $x \leq 0$. Hence $P_{2k}(x) > 0$ for all x.

1975

Problem 1.

Let a, b and c be real numbers such that $a > c \geq 0$, $b > 0$ and

$$ab^2 \left(\frac{1}{(a+c)^2} + \frac{1}{(a-c)^2} \right) = a - b.$$

Find a simple relation among a, b and c.

Solution:

Note that $\dfrac{1}{(a+c)^2} + \dfrac{1}{(a-c)^2} = \dfrac{2(a^2+c^2)}{(a^2-c^2)^2}$. The given relation may be rewritten as $2a(a^2+c^2)b^2 + (a^2-c^2)^2 b - a(a^2-c^2)^2 = 0$, which is a quadratic equation in b. The square of its discriminant is

$$(a^2 - c^2)^4 + 8a^2(a^2 - c^2)^2(a^2 + c^2)$$
$$= (a^2 - c^2)^2(a^4 - 2a^2c^2 + c^4 + 8a^4 + 8a^2c^2)$$
$$= (a^2 - c^2)^2(3a^2 + c^2)^2.$$

It follows that $b = \dfrac{-(a^2-c^2)^2 \pm (a^2-c^2)(3a^2+c^2)}{4a(a^2+c^2)}$. Since we have $a > c \geq 0$ and $b > 0$, we take the plus sign, whereby the numerator becomes

$$(a^2 - c^2)(3a^2 + c^2 - a^2 + c^2) = 2(a^2 - c^2)(a^2 + c^2).$$

Hence $b = \dfrac{a^2 - c^2}{2a}$ so that a simple relation is $2ab = a^2 - c^2$.

Problem 2.

A quadrilateral is inscribed in a convex polygon. Is it always possible to inscribe in this polygon a rhombus whose side is not shorter than the shortest side of the quadrilateral?

Solution:
This is not always possible, and we construct a counterexample. Let $ABCD$ be a kite with $DA = AB < BC = CD$ such that $\angle BAD > \angle BCD > 120°$. $ABCD$ will serve as our overall polygon and as the inscribed quadrilateral. Note that $AB > AC$ since $\angle ABC$ is the smallest angle of triangle ABC. Let $EFGH$ be a rhombus inscribed in $ABCD$. We shall prove that $EF < AB$. We consider two cases.

Case 1. Each of the vertices of $EFGH$ lies on a different side of $ABCD$.
We may assume that E, F, G and H lie respectively on AB, BC, CD and DA. Then $\angle BEF + \angle DHG = \angle BAD < 180°$. We may assume that $\angle BEF < 90°$. Then $\angle AEF > 90° > \angle EAF$, so that

$$EF < AF \le \max\{AB, AC\} = AB.$$

Case 2. One of the sides of $ABCD$ contains two vertices of $EFGH$.
Assume first that E and F lie on AB. Then $EF \le AB$. If $EF = AB$, then $EH \le \max\{AC, DA\} = DA = AB = EF$, so that H must coincide with D. Now G is an interior point of $ABCD$, and $EFGH$ is not an inscribed rhombus. Assume now that F and G lie on BC. If E lies on AB, then F does not coincide with B and we have $EF \le AF < AB$ as before. Hence E does not lie on AB, and H must lie on CD in order for $EFGH$ to be an inscribed rhombus. Clearly, $\angle AEH > 90°$, and we have $EF \le AH < AD = AB$.

Problem 3.
Let the sequence $\{x_n\}$ be defined by $x_0 = 5$ and $x_{n+1} = x_n + \frac{1}{x_n}$ for $n \ge 0$. Prove that $45 < x_{1000} < 45.1$.

Solution:

For $n \geq 0$, $x_{n+1}^2 > x_n^2 + 2$. It follows that

$$x_{1000}^2 > x_{999}^2 + 2 > x_{998}^2 + 4 > \cdots > x_0^2 + 2000 = 2025.$$

Hence $x_{1000} > 45$. We now establish the upper bound. Since $x_n \geq 5$ holds for all $n \geq 0$, $x_{n+1} \leq x_n^2 + 2 + \frac{1}{25} = x_n^2 + \frac{51}{25}$. Now

$$x_{100}^2 \leq a_{99}^2 + \frac{51}{25} \leq x_{98}^2 + \frac{2 \times 51}{25} \leq \cdots \leq x_0^2 + \frac{100 \times 51}{25} = 229.$$

On the other hand,

$$x_{100}^2 > x_{99}^2 + 2 > x_{98}^2 + 4 > \cdots > x_0^2 + 200 = 225.$$

Hence $x_{100} > 15$, so that $x_n > 15$ holds for all $n \geq 100$. It follows that we have $x_{n+1}^2 < x_n^2 + 2 + \frac{1}{225} = x_n^2 + \frac{451}{225}$, so that

$$x_{1000}^2 < x_{999}^2 + \frac{451}{225} < x_{998}^2 + \frac{2 \times 451}{225} < \cdots < a_{100}^2 + \frac{900 \times 451}{225}$$

$$\leq 229 + 1804 = 2033 < 45.1^2.$$

Hence $x_{1000} < 45.1$.

1976

Problem 1.

P is a point outside a parallelogram $ABCD$ such that $\angle PAB$ and $\angle PCB$ are equal but have opposite orientations. Prove that $\angle APB = \angle DPC$.

Solution:

Complete the parallelogram $BAPQ$. Then $\angle PQB = \angle PAB$ and these two angles have opposite orientations. Hence $CBPQ$ is cyclic. Now $CDPQ$ is also a parallelogram. It follows that $\angle DPC = \angle PCQ = \angle PBQ = \angle APB$.

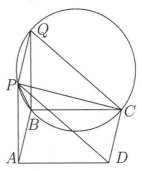

Problem 2.

In 5^5 distinct subsets of size five of the set $\{1,2,3,\ldots,90\}$, if every two of them have at least one common number, prove that there exist four numbers such that each of the 5^5 subsets contains at least one of these four numbers.

Solution:

Suppose that no such subset of four numbers exist. Let A be one of the subsets. Then each of the other $5^5 - 1$ subsets contains one of its five numbers. Suppose the number of subsets containing a particular number a in A is not less than those containing any of the other four. Then at least $5^4 + 1$ subsets contains a, including A. Now there is at least one subset B which does not contain a. Among the $5^4 + 1$ subsets which do contain a, suppose the number of those containing a particular number b in B is not less than those containing any of the other four. Then at least $5^3 + 1$ subsets contain both a and b. Similarly, there exists a number c such that at least $5^2 + 1$ subsets contain all of a, b and c, a number d such that at least $5^1 + 1$ subsets contain all of a, b, c and d, and finally a number such that at least $5^0 + 1 = 2$ subsets contain all of a, b, c, d and e. We have a contradiction since these two subsets are not distinct.

Problem 3.

Prove that if $ax^2 + bx + c > 0$ for all real number x, then we can express $ax^2 + bx + c$ as the quotient of two polynomials whose coefficients are all positive.

Solution:

Since $ax^2 + bx + c$ takes only positive values, we must have $c > 0$ when $x = 0$, and $a > 0$ when x is sufficiently large. If $b > 0$ as well, we have $x^2 + ax + b = \frac{f(x)}{g(x)}$ where $f(x) = x^2 + ax + b$ and $g(x) = 1$. Suppose $b < 0$. Now we must have $b^2 - 4ac < 0$ as otherwise $x^2 + a + b$ has real roots. Choose real numbers λ and mu such that $\lambda > -\frac{b}{c} > 0$ and $1 > \mu > \frac{b^2}{4ac} > 0$. With λ and μ fixed, choose a positive integer n such that $a + b\lambda\mu^{n-1} > 0$. Such an n exists since $\mu < 1$. Define $f(x) = \sum_{k=0}^{n} \lambda^k \mu^{\binom{k}{2}} x^k$. All its coefficients are positive. Let $g(x) = (ax^2 + bx + c)f(x)$. Then we have $ax^2 + bx + c = \frac{f(x)}{g(x)}$.

We claim that all the coefficients of $g(x)$ are also positive. We have

$$
\begin{aligned}
g(x) &= a\lambda^n \mu^{\binom{n}{2}} x^n + \left(a\lambda^{n-1}\mu^{\binom{n-1}{2}} + b\lambda^n \mu^{\binom{n}{2}}\right)x^{n-1} \\
&\quad + \sum_{k=2}^{n-2}\left(a\lambda^{k-2}\mu^{\binom{k-2}{2}} + b\lambda^{k-1}\mu^{\binom{k-1}{2}} + c\lambda^k \mu^{\binom{k}{2}}\right)x^k \\
&\quad + (b + \lambda c)x + c \\
&= a\lambda^n \mu^{\binom{n}{2}} x^n + \lambda^{n-1}\mu^{\binom{n-1}{2}}\left(a + b\lambda\mu^{n-1}\right)x^{n-1} \\
&\quad + \sum_{k=2}^{n-2}\lambda^{k-2}\mu^{\binom{k-2}{2}}\left(a + b\lambda\mu^{k-2} + c\lambda^2\mu^{2k-3}\right)x^k \\
&\quad + (b + \lambda c)x + c.
\end{aligned}
$$

The first and the last coefficients are positive. The second is positive by the choice of n and the second last is positive by the choice of λ. For $2 \leq k \leq n - 2$,

$$
a + c\lambda^2\mu^{2k-3} \geq 2\sqrt{ac\lambda^2\mu^{2k-3}} = \lambda\mu^{k-2}\sqrt{4ac\mu} > -b\lambda\mu^{k-2}
$$

by the choice of μ. This justifies the claim.

1977

Problem 1.
Prove that for any integer $n \geq 2$, the number $n^4 + 4^n$ cannot be prime.

Solution:
Suppose $n^4 + 4^n$ is a prime number greater than 5. Then n must be an odd integer $2k + 1$ with $k \geq 1$. Now

$$
\begin{aligned}
n^4 + 4^n &= (2k+1)^4 + 4^{2k+1} \\
&= ((2k+1)^2)^2 + (2^{2k+1})^2 + 2(2k+1)^2 2^{2k+1} \\
&\quad -2(2k+1)^2 2^{2k+1} \\
&= ((2k+1)^2 + 2^{2k+1})^2 - ((2k+1)2^{k+1})^2 \\
&= ((2k+1)^2 + 2^{2k+1} + (2k+1)2^{k+1}) \\
&\quad ((2k+1)^2 + 2^{2k+1} - (2k+1)2^{k+1}).
\end{aligned}
$$

Since $n^4 + 4^n$ is prime, we must have

$$
(2k+1)^2 + 2^{2k+1} - (2k+1)2^{k+1} = (2k+1-2^k)^2 + 2^{2k} = 1.
$$

However, this is only possible if $k = 0$. We have a contradiction.

Problem 2.
H is the orthocenter of triangle ABC. The medians from A, B and C intersect the circumcircle at D, E and F respectively. P, Q and R are the images of D, E and F under 180° rotations about the midpoints of BC, CA and AB respectively. Prove that H lies on the circumcircle of triangle PQR.

Solution:
Let G be the centroid of ABC, L be the midpoint of BC and K be the foot of the altitude from A. Extend HK to M and HL to N so that $HK = KM$ and $HL = LN$. Then MN is parallel to KL and hence perpendicular to AH.

We claim that both M and N lie on the circumcircle of ABC. Since MBK is congruent to HBK, it is similar to CAK. Hence M lies on the circumcircle. N also lies on the circumcircle since $LN = LM$. It follows that AN is a diameter of the circumcircle, so that $\angle ADN = 90°$. Since PHL is congruent to DNL, $\angle GPH = 90°$ so that P lies on the circle with diameter GH. By symmetry, Q and R also lie on this circle.

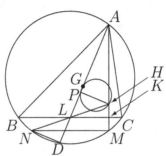

Problem 3.

Let n be a positive integer. There are n students in each of three schools. Each student knows a total of $n + 1$ students in the other two schools. Prove that there exist three students from different schools who know one another.

Solution:

Note that each student knows at least one student in each of the other two schools. Let X be a student who knows $k \geq 1$ students in another school, and k is the smallest such value. By symmetry, we may assume that X is in school A, knows k students in school B and $n + 1 - k \geq k$ students in school C. Let Y be any of those k students in school B. We claim that Y must know a student Z in school C who knows X. If not, then Y knows at most $n - (n + 1 - k) = k - 1$ students in school C. This contradicts the minimality assumption on X.

1978

Problem 1.

Let a and b be rational numbers. Prove that if the equation $ax^2 + by^2 = 1$ has a rational solution (x, y), then it has infinitely many rational solutions.

Solution:

For any rational number k, $\left(\dfrac{k^2 - ab}{k^2 + ab}, \dfrac{2k}{k^2 + ab} \right)$ is a solution of $x^2 + aby^2 = 1$ since $(k^2 - ab)^2 + (2k)^2 = (k^2 + ab)^2$. Moreover, different values of k yield different solutions. Let (u, v) be a solution of $ax^2 + by^2 = 1$. For any solution (x, y) of $x^2 + aby^2 = 1$, $(ux + bvy, vx - auy)$ is a solution of $ax^2 + by^2 = 1$ since we have $a(ux + bvy)^2 + b(vx - auy)^2 = (x^2 + aby^2)(au^2 + bv^2) = 1$. Let (x_1, y_1) and (x_2, y_2) be two different solutions of $x^2 + aby^2 = 1$. We may assume that $y_1 \neq y_2$. Suppose $ux_1 + bvy_1 = ux_2 + bvy_2$ and $vx_1 - auy_1 = vx_2 - auy_2$. Then $\dfrac{bv}{u} = \dfrac{x_1 - x_2}{y_2 - y_1} = \dfrac{-au}{v}$ so that $au^2 + bv^2 = 0$. This is a contradiction. It follows that $ax^2 + by^2 = 1$ has infinitely many rational solutions.

Problem 2.

Let n be an odd positive integer. The vertices of a convex n-gon are painted so that adjacent vertices have different colors. Prove that the polygon can be divided into triangles by non-intersecting diagonals such that none of these diagonals has its endpoints painted in the same color.

Solution:
Since n is odd, $n = 2k - 1$ for some $k \geq 2$. We use mathematical induction on k. For $k = 2$, the polygon itself is a triangle. There are no diagonals and the result holds trivially. Suppose it holds for some $k \geq 2$. Consider a convex $(2k + 1)$-gon $A_1 A_2 \ldots A_{2k+1}$ with vertices painted so that adjacent vertices have different colors. Consider every diagonal which skips over exactly one vertex. If each joins two vertices of the same color, then every other vertex has the same color. Since the number of vertices is odd, they all have the same color. This is a contradiction. It follows that there exists such a diagonal joining two vertices of different colors. Let this diagonal be $A_2 A_4$. We consider three cases.

Case 1. A_1 and A_4 have different colors.
We can use the diagonals $A_2 A_4$ and $A_1 A_4$ to cut off two triangles, leaving behind a convex $(2k - 1)$-gon to which the induction hypothesis can be applied.

Case 2. A_2 and A_5 have different colors.
The argument is analogous to that in Case 1.

Case 3. A_1 and A_4 have the same color, as do A_2 and A_5.
Since the color of A_3 is different from those of A_2 and A_4, it is also different from those of A_1 and A_5. We can use the diagonals $A_1 A_3$ and $A_3 A_5$ to cut off two triangles, leaving behind a convex $(2k - 1)$-gon to which the induction hypothesis can be applied.

Problem 3.
In a triangle with no obtuse angles, r is the inradius, R is the circumradius and H is the longest altitude. Prove that we have $H \geq R + r$.

Solution:
Let ABC be the triangle with $a = BC$, $b = CA$, $c = AB$, $\alpha = \angle A$, $\beta = \angle B$ and $\gamma = \angle C$. Let x, y and z be the respective distances of the circumcenter O from BC, CA and AB. Let D be the foot of perpendicular from O to BC.

By the Law of Sines, $BD = \frac{a}{2} = R\sin\alpha$. It follows from Pythagoras' Theorem that

$$x = OD = \sqrt{OB^2 - BD^2} = R\sqrt{1 - \sin^2\alpha} = R\cos\alpha.$$

Hence

$$\begin{aligned}
x + y + z &= R(\cos\alpha + \cos\beta + \cos\gamma) \\
&= R\left(2\cos\frac{\alpha+\beta}{2}\cos\frac{\alpha-\beta}{2} - \cos(\alpha+\beta)\right) \\
&= R\left(2\cos\frac{\alpha+\beta}{2}\cos\frac{\alpha-\beta}{2} - 2\cos^2\frac{\alpha+\beta}{2} + 1\right) \\
&= R\left(2\cos\frac{\alpha+\beta}{2}\left(\cos\frac{\alpha-\beta}{2} - \cos\frac{\alpha+\beta}{2}\right) + 1\right) \\
&= R\left(2\sin\frac{\gamma}{2}\left(2\sin\frac{\alpha}{2}\sin\frac{\beta}{2}\right) + 1\right) \\
&= 4R\sin\frac{\alpha}{2}\sin\frac{\beta}{2}\sin\frac{\gamma}{2} + R.
\end{aligned}$$

Let F be the point of tangency of the incircle with AB. Then

$$\begin{aligned}
4R\sin\frac{\gamma}{2}\cos\frac{\gamma}{2} &= 2R\sin\gamma \\
&= c \\
&= AF + BF \\
&= r\left(\cot\frac{\alpha}{2} + \cot\frac{\beta}{2}\right) \\
&= r\frac{\cos\frac{\alpha}{2}\sin\frac{\beta}{2} + \cos\frac{\beta}{2}\sin\frac{\alpha}{2}}{\sin\frac{\alpha}{2}\sin\frac{\beta}{2}} \\
&= r\frac{\sin\frac{\alpha+\beta}{2}}{\sin\frac{\alpha}{2}\sin\frac{\beta}{2}} \\
&= r\frac{\cos\frac{\gamma}{2}}{\sin\frac{\alpha}{2}\sin\frac{\beta}{2}}.
\end{aligned}$$

Cancelling $\cos\frac{\gamma}{2}$, we have $r = 4R\sin\frac{\alpha}{2}\sin\frac{\beta}{2}\sin\frac{\gamma}{2}$, so that $x + y + z = r + R$. We may assume that $a \geq b \geq c$ so that the altitude H on c is longest. Twice the area of ABC is given by $cH = ax + by + cz \geq c(x + y + z) = c(r + R)$, so that $H \geq r + R$. Hence equality holds if $a = b = c$. There is another case of equality when $\alpha = 90°$. Then $x = 0$, and the condition is $b = c$. In summary, $H \geq r + R$, and equality holds if and only if ABC is an equilateral or a right isosceles triangle.

1979

Problem 1.

A convex pyramid has an odd number of lateral edges of equal length, and the dihedral angles between neighboring faces are all equal. Prove that the base is a regular polygon and determine when equality holds.

Solution:

Let the apex of the pyramid be V and its base be a convex $(2n+1)$-gon $P_1 P_2 \ldots P_{2n+1}$. Then $VP_1 = VP_2 = \cdots = VP_{2n+1}$. Let O be the foot of perpendicular from V onto the base. For $1 \leq i \leq 2n+1$, $OP_i^2 = VP_i^2 - OV^2$ is constant. Hence O is the circumcenter of $P_1 P_2 \ldots P_{2n+1}$. Now $VP_2 P_3$ is an isosceles triangle symmetric about the plane through V, O and the midpoint M of $P_2 P_2$. Since the dihedral angle between $VP_2 P_3$ and $VP_1 P_2$ is equal to the dihedral angle between $VP_2 P_3$ and $VP_3 P_4$, P_1 and P_4 must also be symmetric about VOM. It follows that $P_1 P_2 = P_3 P_4$. In an analogous manner, we can prove that every other edge of the base has the same length. Since the base has an odd number of edges, it is an equilateral polygon. Since it has a circumcircle, it is regular.

Problem 2.

A real-valued function f defined on all real numbers is such that $f(x) \leq x$ for all real number x and $f(x+y) \leq f(x) + f(y)$ for all real numbers x and y. Prove that $f(x) = x$ for all real number x.

Solution:

From $f(x) \leq x$, we have $f(0) \leq 0$. From $f(x+y) \leq f(x) + f(y)$, we have $f(0) = f(0+0) \leq f(0) + f(0) = 2f(0)$, which implies that $f(0) \geq 0$. It follows that $f(0) = 0$. Setting $y = -x$, we have

$$0 = f(0) = f(x + (-x)) \leq f(x) + f(-x) \leq x + (-x) = 0.$$

Hence we must have $f(x) = x$ for all real number x.

Problem 3.

Let $n \geq 2$ be an integer. An $n \times n$ table of letters is such that no two rows are identical. Prove that a column may be deleted such that in the resulting $n \times (n - 1)$ table, no two rows are identical.

Solution:

More generally, we claim that for any $n \times k$ table with $k \geq n \geq 2$ such that no two rows are identical, we can delete $k - n + 1$ columns and leave behind an $n \times (n - 1)$ table still with no two rows identical. We use induction on n. For $n = 2$, the two rows are not identical. Hence there exists a column consisting of different letters. Deleting the other $k - 2 + 1 = k - 1$ columns will leave behind a 2×1 table with the desired property. Suppose the result holds for some $n \geq 2$. Consider now an $(n + 1) \times k$ table, $k \geq n + 1$. Ignore the last row for now. By the induction hypothesis, we can delete $k - n + 1$ columns so that in the resulting $n + 1 \times n - 1$ table, no two of the first n rows are identical. If the last row is also different from all other rows, we have the desired result by recovering an arbitrary column. If not, at most one other row is identical to the last row. In the original table, these two rows are not identical. Hence they must differ in a deleted column. If we recover this column, we will have an $(n + 1) \times n$ table with the desired property. This completes the inductive argument.

1980

Problem 1.

The points of space are painted in five colors and there is at least one point of each color. Prove that there exists a plane containing four points of different colors.

Solution:

Suppose A, B, C, D and E are five points of different colors. If any four of them are coplanar, there is nothing further to prove. Hence we may assume that $ABCD$ is a tetrahedron. Consider the line AE. We consider two cases.

Case 1. AE intersects the plane BCD at a point F.

If the color of F is different from those of B, C and D, then the plane $BCDF$ satisfies the requirement. If not, then F is of the same color as say B. Then the plane $ACEF$ satisfies the requirement.

Case 2. AE is parallel to the plane BCD.

Then AE is parallel to at most one of BC, CD and DB. We may assume by symmetry that AE is not parallel to CD. Draw a line through B parallel to AE, intersecting CD at G. If the color of G is different from those of B, C and D, then the plane $BCDG$ satisfies the requirement. If G is of the same color as B, then the plane $ACDG$ satisfies the requirement. Otherwise, G is of the same color as say C. Since BG is parallel to AE, $AEBG$ is a plane which satisfies the requirement.

Problem 2.

Let $n > 1$ be an odd integer. Prove that there exist positive integers x and y such that $\frac{4}{n} = \frac{1}{x} + \frac{1}{y}$ if and only if n has a prime divisor of the form $4k - 1$.

Solution:

We first assume that there exist positive integers x and y which satisfy $\frac{1}{n} = \frac{1}{4x} + \frac{1}{4y}$. Then $4x > n$ and $4y > n$, and we have $(4x - n)(4y - n) = n^2$. If n itself is of the form $4k - 1$, then it cannot have only prime divisors of the form $4k + 1$. If n is of the form $4k + 1$, then $4x - n$ is of the form $4k - 1$ and has a prime divisor of the same form. Since this prime divides n^2, it must divide n. Conversely, suppose n has a prime divisor p of the form $4k - 1$. If n is of the form $4k + 1$, then $p + n$ is divisible by 4. Hence both $x = \frac{p+n}{4}$ and $y = \frac{n(p+n)}{4p}$ are integers. We have $\frac{1}{4x} + \frac{1}{4y} = \frac{1}{p+n} + \frac{p}{n(p+n)} = \frac{1}{n}$. If n is of the form $4k - 1$, we may simply take $x = \frac{n(n+1)}{4}$ and $y = \frac{n+1}{4}$. Then $\frac{1}{4x} + \frac{1}{4y} = \frac{1}{n(n+1)} + \frac{1}{n+1} = \frac{1}{n}$.

Problem 3.

Two tennis clubs have 1000 and 1001 members respectively. All 2001 players have different strength, and in a match between two players, the stronger one always wins. The ranking of the players within each club is known. Find a procedure using at most 11 games to determine the 1001st players in the total ranking of the 2001 players.

Solution:

More generally, let there be m members in the first club and n members in the second, where $1 \leq m \leq n \leq m + 1$. Let k be the smallest integer such that $n \leq 2^k$. We use simultaneous induction on k to prove that $k + 1$ games are sufficient to find

(1) the $(m + 1)$st player when $n = m + 1$;

(2) the mth player when $n = m$;

(3) the $(m + 1)$st player when $n = m$.

When $n = 1001$, $k = 10$. By (1), the task can be accomplished in 11 games as desired. For $k = 0$, $m = n = 1$, and a single game furnishes a complete ranking of the two players. Suppose the results hold for some $k \geq 0$. Consider n where $2^k < n + 1 \leq 2^{k+1}$. We consider two cases for (1).

Case 1. $m = 2h$ for some positive integer h.

Let the members in the first club be A_1, A_2, ..., A_{2h} with respective strengths $a_1 > a_2 > \cdots > a_{2h}$. Let the members in the second club be B_1, B_2, ..., B_{2h+1} with respective strengths $b_1 > b_1 > \cdots > b_{2h+1}$. Use a game to compare a_h and b_{h+1}. Suppose $a_h > b_{h+1}$. Then each of A_1, A_2, ..., A_h is ahead of all of A_{h+1}, A_{h+2}, ..., A_{2h}, B_{h+1}, B_{h+2}, ..., B_{2h+1}. Hence they are among the top $2h$ players. On the other hand, each of B_{h+2}, B_{h+3}, ..., B_{2h+1} is behind all of B_1, B_2, ..., B_{h+1}, A_1, A_2, ..., A_h. Hence they are among the bottom $2h$ players. Kicking out these $2h$ players from the clubs leave behind h members in the first club and $h + 1$ players in the second, with $2^{k-1} < h \leq 2^k$. By the induction hypothesis, the $(h + 1)$st player among the members in the reduced clubs can be found in $k + 1$ more games, and this player is be the $(2h + 1)$st player in the original clubs. Hence $k + 2$ games are sufficient. Suppose $a_h < b_{h+1}$. Then each of B_1, B_2, ..., B_{h+1} is ahead of all of A_h, A_{h+1}, ..., A_{2h}, B_{h+2}, B_{h+3}, ..., B_{2h+1}. Hence they are among the top $2h$ players. On the other hand, each of A_{h+1}, A_{h+2}, ..., A_{2h} is behind all of A_1, A_2, ..., A_h, B_1, B_2, ..., B_{h+1}. Hence they are among the bottom $2h$ players. This can be handled exactly as before by kicking out these $2h$ players.

Case 2. $m = 2h - 1$ for some positive integer h.
Let the members in the first club be A_1, A_2, ..., A_{2h-1} with respective strengths $a_1 > a_2 > \cdots > a_{2h-1}$. Let the members in the second club be B_1, B_2, ..., B_{2h} with respective strengths $b_1 > b_1 > \cdots > b_{2h}$. Use a game to compare a_h and b_{h+1}. Suppose $a_h > b_{h+1}$. Then each of A_1, A_2, ..., A_{h-1} is ahead of all of A_h, A_{h+1}, ..., A_{2h-1}, B_{h+1}, B_{h+2}, ..., B_{2h}. Hence they are among the top $2h - 1$ players. On the other hand, each of B_{h+1}, B_{h+2}, ..., B_{2h} is behind all of B_1, B_2, ..., B_h, A_1, A_2, ..., A_h. Hence they are among the bottom $2h - 1$ players. Kicking out these $2h - 1$ players from the clubs leave behind h members in each club, with $2^{k-1} < h \le 2^k$. By the induction hypothesis in (3), the $(h+1)$st player among the members in the reduced clubs can be found in $k + 1$ more games, and this player will also be the $(2h)$th player in the original clubs. Hence $k + 2$ games are sufficient. Suppose $a_h < b_{h+1}$. Then each of B_1, B_2, ..., B_h is ahead of all of A_h, A_{h+1}, ..., A_{2h}, B_{h+1}, B_{h+2}, ..., B_{2h+1}. Hence they are among the top $2h - 1$ players. On the other hand, each of A_{h+1}, A_{h+2}, ..., A_{2h-1} is behind all of A_1, A_2, ..., A_h, B_1, B_2, ..., B_{h+1}. Hence they are among the bottom $2h - 1$ players. This can be handled as before by kicking out these $2h - 1$ players, using the induction hypothesis in (2).
We can handle (2) and (3), which are equivalent, in a similar manner.

1981

Problem 1.
Prove that any five points A, B, P, Q and R in a plane,

$$AB + PQ + QR + RP \leq AP + AQ + AR + BP + BQ + BR.$$

Solution:
We first assume that two of P, Q and R, say P and Q, are on the same side of AB while the third point, R, is on the other side. We consider two cases.
Case 1. A, B, P and Q determine a convex quadrilateral.
Let it be $ABPQ$, and let C be the point of intersection of AP and BQ. By the Triangle Inequality, we have $AQ + AR \geq QR$, $BR + BP \geq RP$ and

$$AP + BQ = AC + CP + BC + CQ \geq AB + PQ.$$

Hence $AP + AQ + AR + BP + BQ + BR \geq AB + PQ + QR + RP$.
Case 2. A, B, P and Q determine a non-convex quadrilateral.
Let Q be inside ABP. We may assume that R lies on the same side of PQ as B. Then Q is also inside PAR. It follows that $AP + AR \geq PQ + QR$. Along with $AQ + BQ \geq AB$ and $BR + BP \geq RP$, we have the desired result.

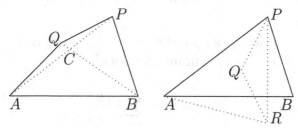

We now assume that all of P, Q and R are on the same side of AB. If the convex hull of A and B with two of P, Q and R is a quadrilateral, we can use the argument in Case 1. Suppose all three are triangles. We may assume that P is the farthest from AB. Then Q and R are both inside PAB. Moreover, the line QR must intersect one of the segments AP or BP, say AP. Then Q is inside PAR and we can use the argument in Case 2.

Problem 2.
Let $n > 2$ be an even integer. The squares of an $n \times n$ chessboard are painted with $\frac{1}{2}n^2$ colors, with exactly two squares of each color. Prove that n rooks can be placed on n squares of different colors such that no two of them attack each other.

Solution:
There are $n!$ sets of n squares with no two in the same row and no two in the same column. For $1 \le k \le \frac{1}{2}n^2$, the number of such sets of squares containing both squares of color k is at most $(n-2)!$. Since $n > 2$, $\frac{1}{2}n^2(n-2)!$ is less than $n!$. Hence there is at least one such set of squares not containing two squares of the same color.

Problem 3.
For a positive integer n, $r(n)$ denote the sum of the remainders when n is divided by 1, 2, ..., n respectively. Prove that for infinitely many positive integers n, $r(n) = r(n+1)$.

Solution:
For $1 \le i \le n$, let r_i denote the remainder when n is divided by i. Then $n = i\lfloor \frac{n}{i} \rfloor + r_i$. Summation yields

$$r(n) = n^2 - \sum_{i=1}^{n} i \left\lfloor \frac{n}{i} \right\rfloor.$$

Similarly,

$$r(n+1) = (n+1)^2 - \sum_{i=1}^{n} i \left\lfloor \frac{n+1}{i} \right\rfloor.$$

Thus $r(n) - r(n+1) = \sum_{i=1}^{n} i\left(\left\lfloor \frac{n+1}{i} \right\rfloor - \left\lfloor \frac{n}{i} \right\rfloor\right) - (2n+1)$. For $1 \leq i \leq n+1$, $\lfloor \frac{n+1}{i} \rfloor - \lfloor \frac{n}{i} \rfloor = 1$ if i divides $n+1$, and 0 otherwise. When $n + 1 = 2^k$ for some positive integer k,

$$r(n) - r(n+1) = \sum_{j=0}^{k} 2^j - (2n+1) = 2^{k+1} - 1 - (2n+1) = 0.$$

It follows that there are infinitely many values of n, of the form $2^k - 1$, for which $r(n) = r(n+1)$.

1982

Problem 1.

A cube has integer side length and all four vertices of one face are lattice points. Prove that the other four vertices are also lattice points.

Solution:

Let the cube be $ABCD - EFGH$ with edges AE, BF, CG and DH between the bases $ABCD$ and $EFGH$. Let A, B, C and D be lattice points. We may take A to be the origin, and the let the coordinates for B, D and E be (b_1, b_2, b_3), (d_1, d_2, d_3) and (e_1, e_2, e_3) respectively, where b_1, b_2, b_3, d_1, d_2 and d_3 are integers. The desired result will follow if we can show that e_1, e_2 and e_3 are also integers. Let a be the integer length of the cube. Then

$$b_1^2 + b_2^2 + b_3^2 = d_1^2 + d_2^2 + d_3^2 = e_1^2 + e_2^2 + e_3^2 = a^2.$$

Since AB, AD and AE are mutually perpendicular, their pairwise dot products yield

$$b_1 d_1 + b_2 d_2 + b_3 d_3 = b_1 e_1 + b_2 e_2 + b_3 e_3 = d_1 e_1 + d_2 e_2 + d_3 e_3 = 0.$$

Their cross product yields $e_1 = k(b_2 d_3 - b_3 d_2)$, $e_2 = k(b_3 d_1 - b_1 d_3)$ and $e_3 = k(b_1 d_2 - b_2 d_1)$. Now

$$
\begin{aligned}
a^2 &= e_1^2 + e_2^2 + e_3^2 \\
&= k^2 (b_2^2 d_3^2 + b_3^2 d_2^2 + b_3^2 d_1^2 + b_1^2 d_3^2 + b_1^2 d_2^2 + b_2^2 d_1^2 \\
&\quad - 2(b_2 d_2 b_3 d_3 + b_3 d_3 b_1 d_1 + b_1 d_1 b_2 d_2) \\
&= k^2 ((b_1^2 + b_2^2 + b_3^2)(d_1^2 + d_2^2 + d_3^2) - (b_1^2 d_1^2 + b_2^2 d_2^2 + b_3^2 d_3^2) \\
&\quad - 2(b_2 d_2 b_3 d_3 + b_3 d_3 b_1 d_1 + b_1 d_1 b_2 d_2)) \\
&= k^2 ((b_1^2 + b_2^2 + b_3^2)(d_1^2 + d_2^2 + d_3^2) - (b_1 d_1 + b_2 d_2 + b_3 d_3)^2) \\
&= k^2 a^4.
\end{aligned}
$$

It follows that $k = \frac{1}{a}$ so that $e_1 = \frac{b_2d_3 - b_3d_2}{a}$, $e_2 = \frac{b_3d_1 - b_1d_3}{a}$ and $e_3 = \frac{b_1d_2 - b_2d_1}{a}$. We have

$$
\begin{aligned}
(b_2d_3 - b_3d_2)^2 &= b_2^2d_3^2 + b_3^2d_2^2 - 2b_2d_2b_3d_3 \\
&= b_2^2d_3^2 + b_3^2d_2^2 - b_2d_2(-b_1d_1 - b_2d_2) \\
&\quad - b_3d_3(-b_3d_3 - b_2d_1) \\
&= b_2^2d_3^2 + b_3^2d_2^2 + b_2^2d_2^2 + b_3^2d_3^2 + b_1d_1(b_2d_2 + b_3d_3) \\
&= (b_2^2 + b_3^2)(d_2^2 + d_3^2) - b_1^2d_1^2 \\
&= (a^2 - b_1^2)(a^2 - d_1)^2 - b_1^2d_1^2 \\
&= a^2(a^2 - b_1^2 - d_1^2).
\end{aligned}
$$

It follows that $e_1^2 = \left(\frac{b_2d_3 - b_3d_2}{a}\right)^2 = a^2 - b_1^2 - d_1^2$ is an integer. Since e_1 is rational, it is an integer. Similarly, e_2 and e_3 are also integers.

Problem 2.
Prove that for any integer $k > 2$, there exist infinitely many positive integers n such that the least common multiple of n, $n + 1, \ldots, n + k - 1$ is greater than the least common multiple of $n + 1$, $n + 2, \ldots, n + k$.

Solution:
Let $\triangle$ denote greatest common divisors and $\bigtriangledown$ denote least common multiples. Let $n - 1$ be any integral multiple of $k!$. There are infinitely many such values of n, say $n = mk!$ for some positive integer m. For $1 \le i \le k - 1$, $n \triangle i = n \triangle (n + i) = 1$. We also have

$$
n \bigtriangledown (n+1) \bigtriangledown \cdots \bigtriangledown (n+k-1) = n((n+1) \bigtriangledown \cdots \bigtriangledown (n+k-1)).
$$

Now $(n + k) \triangle (n + 1) = (n + k) \triangle (k - 1) = k - 1$ since $k - 1$ divides $n + k = mk! + k - 1$. Hence

$$
(n + k) \triangle ((n+1) \bigtriangledown \cdots \bigtriangledown (n+k-1)) \ge k - 1.
$$

It follows that

$$(n+1) \bigtriangledown (n+2) \bigtriangledown \cdots \bigtriangledown (n+k)$$
$$= ((n+1) \bigtriangledown (n+2) \bigtriangledown \cdots \bigtriangledown (n+k-1)) \bigtriangledown (n+k)$$
$$= \frac{((n+1) \bigtriangledown (n+2) \bigtriangledown \cdots \bigtriangledown (n+k-1))(n+k)}{((n+1) \bigtriangledown (n+2 \bigtriangledown \cdots \bigtriangledown (n+k-1)) \bigtriangleup (n+k)}$$
$$\leq \frac{((n+1) \bigtriangledown (n+2) \bigtriangledown \cdots \bigtriangledown (n+k-1))(n+k)}{k-1}.$$

Since $\frac{n+k}{k-1} \leq \frac{n+k}{2} < n$, we have

$$n \bigtriangledown (n+1) \bigtriangledown \cdots \bigtriangledown (n+k-1) \geq (n+1) \bigtriangledown (n+2) \bigtriangledown \cdots \bigtriangledown (n+k).$$

Problem 3.
The set of integers are painted in 100 colors, with at least one number of each color. For any two intervals $[a, b]$ and $[c, d]$ of equal length and with integral endpoints, if a has the same color as c and b has the same color as d, then $a + x$ has the same color as $c + x$ for any integer x, $0 \leq x \leq b - a$. Prove that the numbers 1982 and -1982 are painted in different colors.

Solution:
For any integer k, the interval $[101k, 101k + 100]$ contains 101 integers. By the Pigeonhole Principle, two of them, say a_k and b_k, must have the same color. We may take $a_k < b_k$ so that $0 < b_k - a_k \leq 100$. Hence $b_k - a_k$ can take only 100 different values. By the infinite version of the Pigeonhole Principle, there exists a value t such that $t = b_k - a_k$ for infinitely many k. We claim that for all integers x, x and $x + t$ have the same color. We can choose integers i and j such that $a_i < x < a_j$. Consider the intervals $[a_i, a_j]$ and $[b_i, b_j]$. Since $b_i - a_i = t = b_j - a_j$, $a_j - a_i = b_j - b_i$, so that the two intervals have equal length. Since a_i and a_j have the same colors as b_i and b_j respectively. Hence the entire intervals have the same color. Since $b_i = a_i + t$ and $b_j = a_j + t$, $x + t$ is in $[b_i, b_j]$ and as the same color as x as claimed.

Hence every t integers have the same color, and $t \leq 100$. Since all 100 colors are used, $t = 100$. It follows that integers x and y have the same color if and only if $x \equiv y \pmod{100}$. Since $1982 \not\equiv -1982 \pmod{100}$, they have different colors.

1983

Problem 1.
Let x, y and z be rational numbers such that

$$x^3 + 3y^3 + 9z^3 - 9xyz = 0.$$

Prove that $x = y = z = 0$.

Solution:
Each term of the equation is of degree 3, so that the equation is still satisfied if we cancel out the common denominator of x, y and z. Thus if there are rational solutions, there will be integer solutions. If there are solutions other than $(0,0,0)$, choose a solution (x, y, z) such that $|x| + |y| + |z|$ is minimum. Clearly, 3 divides x, so that $x = 3u$ for some integer u. Then

$$27u^3 + 3y^3 + 9z^3 - 27uxy = 0$$

so that

$$y^3 + 3z^3 + 9u^3 - 9yzu = 0.$$

Similarly, we have $y = 3v$ and $z = 3w$ for some integers v and w, and they satisfy

$$u^3 + 3v^3 + 9w^3 - 9uvw = 0.$$

Hence (u, v, w) is also an integer solution. However,

$$
\begin{aligned}
|u| + |v| + |w| &= \frac{1}{3}(|x| + |y| + |z|) \\
&< |x| + |y| + |z|.
\end{aligned}
$$

This is a contradiction.

Problem 2.
In the polynomial $f(x) = x^n + a_1 x^{n-1} + \cdots + a_{n-1}x + 1$, n is a positive integer and a_1, $\ldots$, a_{n-1} are non-negative real numbers. Prove that $f(2) \geq 3^n$ if $f(x)$ has n real roots.

Solution:
Since the coefficients of $f(x)$ are all non-negative, $f(x) > 0$ for all real numbers $x > 0$. It follows that its roots are $-t_1$, $-t_2$, ..., $-t_n$ where t_1, t_2, ..., t_n are positive real numbers. Hence $f(x) = (x+t_1)(x+t_2)\cdots(x+t_n)$ so that $t_1 t_2 \cdots t_n = 1$. By the Arithmetic-Geometric Means Inequality,

$$2 + t_k = 1 + 1 + t_k \geq 3\sqrt[3]{t_k}$$

for $1 \leq k \leq n$. It follows that

$$f(2) = (2+t_1)(2+t_2)\cdots(2+t_n) \geq 3^n \sqrt[3]{t_1 t_2 \cdots t_n} = 3^n.$$

Problem 3.
Let n be a positive integer. P_1, P_2, ..., P_n and Q are points in the plane with no three on the same line. For any two different points P_i and P_j, there exists a third point P_k such that Q lies inside triangle $P_i P_j P_k$. Prove that n is odd.

Solution:
From Q, construct rays to P_1, P_2, ..., P_n, and we may assume that they are in clockwise order. For $1 \leq k \leq n$, let QR_k be the ray opposite to QP_k. Consider two adjacent rays QP_i and QP_{i+1}. The ray QR_j must lie between them, where P_j is a point such that Q lies inside triangle $P_i P_{i+1} P_j$. Hence every two adjacent rays contain an opposite ray between them. Since there are n pairs of adjacent rays and n opposite rays, we have a one-to-one correspondence. Consider the rays P_1, P_2, ..., P_k where k is the smallest value such that the rays opposite to P_{k+1}, P_{k+2}, ..., P_n are contained between P_1 and P_k clockwise. Then we have k rays and $k - 1$ opposite rays, and the total $n = k + (k - 1) = 2k + 1$ is an odd number.

1984

Problem 1.
When the first 4 rows of Pascal's Triangle are written down in the usual way and the numbers in vertical columns are added up, we obtain 7 numbers as shown below, 5 of them being odd.

$$
\begin{array}{ccccccc}
 & & & 1 & & & \\
 & & 1 & & 1 & & \\
 & 1 & & 2 & & 1 & \\
1 & & 3 & & 3 & & 1 \\
\hline
1 & 1 & 4 & 3 & 4 & 1 & 1
\end{array}
$$

If this procedure is applied to the first 1024 rows of Pascal's Triangle, how many of the 2047 numbers thus obtained will be odd?

Solution:
Let the 2047 numbers obtained from first 1024 rows of Pascal's Triangle be a_1, a_2, $\ldots$, a_{2047}. We have $a_1 = 1$. For odd values of k, $3 \le k \le 2045$, $a_k = a_{k-1} + a_{k+1}$. This is because every number in the kth column is the sum of the numbers in the $k - 1$st and $k + 1$st columns in the preceding row, if we put an extra 0 on top of the shorter of the $k - 1$st and the $k + 1$st columns. For even values of k, $2 \le k \le 2046$, we claim that $a_k \equiv a_{k-1} + a_{k+1}$ (mod 2). Since $a_1 = a_2 = 1$, the parity pattern of the first 1024 numbers is odd-odd-even-$\cdots$-odd-odd-even-odd. It follows that there are $1024 - (1023 \div 3) = 683$ odd integers among the first 1024 numbers, and the total number of odd integers among the 2047 numbers is $2 \times 683 - 1 = 1365$. We now justify the claim. For even values of k, we have

$$
a_{k-1} + a_{k+1} = a_k + \binom{1023}{\frac{k-1}{2}} + \binom{1023}{\frac{k+1}{2}} = a_k + \binom{1024}{\frac{k+1}{2}}.
$$

Now $\dfrac{k+1}{2}\dbinom{1024}{\frac{k+1}{2}} = 1024\dbinom{1023}{\frac{k-1}{2}}$. Since $k+1 < 2048$, $\dbinom{1024}{\frac{k+1}{2}}$ must be even.

Problem 2.
The rigid plates $A_1B_1A_2$, $B_1A_2B_2$, $A_2B_2A_3$, ..., $B_{13}A_{13}B_{14}$, $A_{14}B_{14}A_1$ and $B_{14}A_1B_1$ are in the shape of equilateral triangles such that they can be folded along common edges A_1B_1, B_1A_2, ..., $A_{14}B_{14}$ and $B_{14}A_1$. Can they be folded so that all 28 plates lie in the same plane?

Solution:
Suppose the plates can lie in the same plane. Mark off an isogonal lattice so that when a plate is placed in the plane, its three vertices are lattice points. Paint the lattice points in three colors as shown in the diagram. When a plate is placed on the lattice, all three vertices are lattice points of different colors. Now A_1 and B_2 must have the same color, being different from those of B_1 and A_2. In turns, this forces A_4, B_5, A_7, B_8, A_{10}, B_{11}, A_{13} and B_{14} to have the same color. However, A_1 and B_{14} are vertices of the same plate. This is a contradiction.

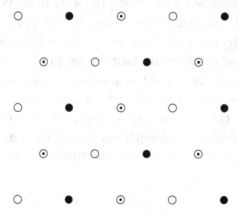

Problem 3.

Let p and q be positive integers. In a set of $n > 1$ integers, if there are two equal ones among them, add p to one and subtract q from the other. Prove that after a finite number of steps, all n numbers are distinct.

Solution:

We shall show that for any positive integer k, there exists a positive integer a_k such that a_k steps are sufficient to make all numbers distinct in a set containing at most k numbers. Clearly, $a_1 = 0$ and $a_2 = 1$. Suppose a_k exists for some $k \geq 2$ but a_{k+1} does not. Then there exists a set of $k + 1$ numbers which cannot be made distinct in a finite number of steps. We call $a_k + 1$ consecutive steps a cycle. We claim that after a cycle, the largest number must increase by at least p. Suppose this number is not involved in any steps in the cycle. We can use the first a_k steps to make the other k numbers distinct. Since the task cannot be accomplished, one of them must be equal to m. In the last step, we must change the two copies of m into $m + p$ and $m - q$. This justifies the claim. Now m can increase by at most p in one step. Hence it can increase by at most $p(a_k + 1)$ in a cycle. Similarly, the smallest number can decrease by at most $q(a_k + 1)$ in a cycle. Choose an integer n such that $pn > k(p + q)(a_k + 1)$. We can make the largest number increase by pn in n cycles. Hence the gap between the largest and the smallest numbers is now greater than $k(p + q)(a_k + 1)$. Consider the k gaps between these $k + 1$ numbers when they are arranged in order. By the Pigeonhole Principle, some gap has length exceeding $(p + q)(a_k + 1)$. Let the end points of this gap be the numbers $u < v$.

Let U be the subset of the $k+1$ numbers less than or equal to u, and V be the subset of the $k+1$ numbers greater than or equal to v. By the definition of a_k, we can make the numbers in U distinct in at most a_k steps. The largest number has increased by at most $p(a_k + 1)$. In another a_k moves, we can make the numbers in V distinct, and the smallest integer has decreased by at most $q(a_k + 1)$. Since the gap between u and v exceeds $(p + q)(a_k + 1)$, all $k + 1$ numbers are now distinct. This is a contradiction. Hence a_{k+1} must exist, and the desired result follows from mathematical induction.

1985

Problem 1.

Let n be a positive integer. The convex $(n+1)$-gon $P_0 P_1 \ldots P_n$ is divided by non-intersecting diagonals into $n-1$ triangles. Prove that these triangles can be numbered from 1 to $n-1$ such that P_i is a vertex of the triangle numbered i for $1 \leq i \leq n-1$.

Solution:

We use mathematical induction on n. For $n = 2$, the only triangle $P_0 P_1 P_2$ in the partition must be labeled 1, and P_1 is indeed a vertex of this triangle. Suppose the result holds for all values up to some $n \geq 2$. Consider $P_0 P_1 \ldots P_{n+1}$. Not both $P_0 P_1 P_{n+1}$ and $P_0 P_n P_{n+1}$ can be triangles in the same partition, there must be a triangle $P_{k-1} P_k P_{k+1}$ in the partition, where $1 \leq k \leq n$. Label it k. By the induction hypothesis, the convex n-gon $P_0 P_1 \ldots P_{k-1} P_{k+1} \ldots P_{n+1}$ may be labeled 1, 2, $\ldots$, $k-1$, $k+1$, $\ldots$, $n+1$ with the desired property. This competes the inductive argument.

Problem 2.

Let n be a positive integer. For each prime divisor p of n, consider the highest power of p which does not exceed n. The sum of these powers is defined as the power-sum of n. Prove that there exist infinitely many positive integers which are less than their respective power-sums.

Solution:
Let p be any odd prime and let $n = 2p$. Let k be the positive integer such that $2^k \le n < 2^{k+1}$. Since $n < p^2$, the power-sum of n is $p + 2^k > p + \frac{n}{2} = n$. Since there are infinitely many odd primes, there are infinitely many n less than their respective power-sums.

Problem 3.
Each vertex of a triangle is reflected across the opposite side. Prove that the area of the triangle determined by the three points of reflection is less than five times the area of the original triangle.

Solution:
Let the triangle be ABC. Let $BC = a$, $CA = b$, $AB = c$, $\angle A = \alpha$, $\angle B = \beta$ and $\angle C = \gamma$. We will take $[ABC] = 1$. Let D, E and F be the respective reflectional images of A, B and C. Then $[BCD] = [CAE] = [ABF] = 1$.

For acute triangles, we have

$$[DEF] = [ABC] + [BCD] + [CAE] + [ABF]$$
$$\pm [AEF] \pm [BFD] \pm [CDE].$$

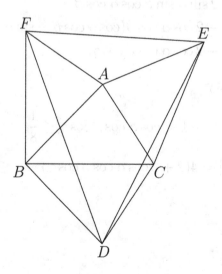

In the diagram above, $\alpha > 60°$. Hence we take the plus sign and $[AEF] = \frac{1}{2}bc\sin(360° - 3\alpha) = -\frac{1}{2}bc\sin 3\alpha$. Since $\beta < 60°$, we take the minus sign and $-[BFD] = -\frac{1}{2}ca\sin 3\beta$. Whether $\gamma > 60°$ or $\gamma < 60°$, the contribution from the term $\pm[CDE]$ is equal to $-\frac{1}{2}ab\sin 3\gamma$. Hence

$$[DEF] = 4 - \frac{1}{2}(bc\sin 3\alpha + ca\sin 3\beta + ab\sin 3\gamma).$$

Using

$$
\begin{aligned}
\sin 3\theta &= \sin(\theta + 2\theta) \\
&= \sin\theta\cos 2\theta + \cos\theta\sin 2\theta \\
&= \sin\theta(1 - 2\sin^2\theta) + 2\sin\theta(1 - \sin^2\theta) \\
&= \sin\theta(3 - 4\sin^2\theta),
\end{aligned}
$$

we have $[DEF] = 4(\sin^2\alpha + \sin^2\beta + \sin^2\gamma) - 5$. Note that

$$
\begin{aligned}
&\sin^2\alpha + \sin^2\beta + \sin^2\gamma \\
&= 1 - \cos^2\alpha + 1 - \cos^2\beta + \sin^2(\alpha + \beta) \\
&= 2 - \cos^2\alpha(1 - \sin^2\beta) - \cos^2\beta(1 - \sin^2\alpha) \\
&\quad + 2\sin\alpha\sin\beta\cos\alpha\cos\beta \\
&= 2 - 2\cos\alpha\cos\beta(\cos\alpha\cos\beta - \sin\alpha\sin\beta) \\
&= 2 + 2\cos\alpha\cos\beta\cos\gamma.
\end{aligned}
$$

By the inequality

$$-1 < \cos\alpha\cos\beta\cos\gamma \le \frac{1}{8},$$

we have $[DEF] = 4(2 + 2\cos\alpha\cos\beta\cos\gamma) - 5 \le 4 < 5$.

For obtuse triangles, we have

$$
\begin{aligned}
[DEF] &= [AEF] + [BFD] + [CDE] \\
&\quad -[ABC] - [BCD] - [CAE] - [ABF] \\
&= \frac{1}{2}(bc\sin 3\alpha + ca\sin 3\beta + ab\sin 3\gamma) - 4 \\
&= 5 - 4(\sin^2\alpha + \sin^2\beta + \sin^2\gamma) \\
&= 5 - 4(2 + 2\cos\alpha\cos\beta\cos\gamma) \\
&< 5.
\end{aligned}
$$

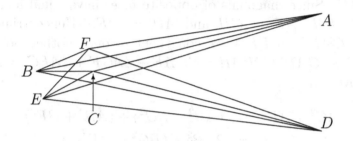

1986

Problem 1.
Prove that three rays from the same point contain three face diagonals of some rectangular block if and only if the rays include pairwise acute angles with sum 180°.

Solution:
We first establish necessity. Consider any rectangular block $ABCD - EFGH$ as shown in the diagram and focus on AC, AF and AH. Since diagonals of opposite faces have equal lengths, $AC = FH$, $AF = CH$ and $AH = CF$. Hence triangles ACF, CAH and FHA are congruent to one another, so that $\angle CAF + \angle CAH + \angle FAH = \angle CAF + \angle ACF + \angle AFC = 180°$. Moreover,

$$
\begin{aligned}
AC^2 + AF^2 &= (AB^2 + BC^2) + (AB^2 + BF^2) \\
&= 2AB^2 + (BC^2 + BF^2) \\
&= 2AB^2 + CF^2 \\
&> AH^2.
\end{aligned}
$$

Hence $\angle CAH$ is acute. We can prove analogously that $\angle CAH$ and FAH are also acute.

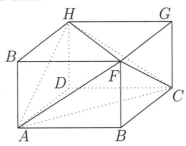

To establish sufficiency, given from the same point three rays which include pairwise acute angles α, β and γ with sum $180°$, we construct a rectangular block with face diagonals contained in them. First construct a triangle with angles α, β and γ, facing sides of lengths x, y and z respectively. Let a, b and c be such that $x = \sqrt{b^2 + c^2}$, $y = \sqrt{c^2 + a^2}$ and $z = \sqrt{a^2 + b^2}$. Then we have $a = \sqrt{\frac{1}{2}(y^2 + z^2 - x^2)}$, $b = \sqrt{\frac{1}{2}(z^2 + x^2 - y^2)}$ and $c = \sqrt{\frac{1}{2}(x^2 + y^2 - z^2)}$. Since α, β and γ are acute, a, b and c are positive real numbers. An $a \times b \times c$ rectangular block will have all the desired properties.

Problem 2.
Let n be any integer greater than 2. Determine the maximum value of h and the minimum value of H such that

$$h < \frac{a_1}{a_1 + a_2} + \frac{a_2}{a_2 + a_3} + \cdots + \frac{a_n}{a_n + a_1} < H$$

for any positive real numbers a_1, a_2, $\ldots$, a_n.

Solution:
Denote $\dfrac{a_1}{a_1 + a_2} + \dfrac{a_2}{a_2 + a_3} + \cdots + \dfrac{a_n}{a_n + a_1}$ by $f(a_1, a_2, \ldots, a_n)$.
Then

$$f(a_1, a_2, \ldots, a_n) + f(a_n, a_{n-1}, \ldots, a_1) = n.$$

This implies that $h + H = n$, so that we need only find h. Note that

$$
\begin{aligned}
& f(a_1, a_2, \ldots, a_n) \\
> \; & \frac{a_1}{a_1 + a_2 + \cdots + a_n} + \frac{a_2}{a_1 + a_2 + \cdots + a_n} \\
& + \cdots + \frac{a_n}{a_1 + a_2 + \cdots + a_n} \\
= \; & 1.
\end{aligned}
$$

Hence $h \geq 1$. Now take $a_i = t^i$ for some positive real number t. Then

$$f(a_1, a_2, \ldots, a_n) = \frac{1}{1+t} + \frac{1}{1+t} + \cdots + \frac{1}{1+t} + \frac{t^n}{t + t^n}.$$

If t is sufficiently large, this expression gets arbitrarily close to 1. Hence we must take $h = 1$, so that $H = n - 1$.

Problem 3.
From the first 100 positive integers, k of them are drawn at random and then added. For which positive integers k is the sum equally likely to be odd or even?

Solution:
First, we claim that the sum equally likely to be odd or even for any odd number k between 1 and 99 inclusive. In any set S of k chosen numbers, there exists a value i, $1 \le i \le 49$, such that exactly one of $2i - 1$ and $2i$ is chosen, and we may assume that this value of i is minimum. Consider now the set T which is identical to S except that $2i - 1$ and $2i$ have been switched. Clearly, one of S and T has odd sum and the other has even sum. This one-to-one correspondence justifies the claim. Suppose k is an even number between 2 and 98 inclusive. The same one-to-one correspondence exists among sets for which there exists a value i, $1 \le i \le 49$, such that exactly one of $2i - 1$ and $2i$ is chosen. Each of the remaining sets of k chosen number consists of pairs of numbers $2i - 1$ and $2i$ for some i, $1 \le i \le 49$. The sum of such a set is even if k is a multiple of 4, and odd otherwise. Hence the sum is not equally likely to be odd or even.

1987

Problem 1.

Find all positive integers a, b, c and d such that $a + b = cd$ and $ab = c + d$.

Solution:

From $a + b + c + d = ab + cd$, we have

$$(a - 1)(b - 1) + (c - 1)(d - 1) = 2.$$

Since a, b, c and d are all positive integers, $(a - 1)(b - 1) \geq 0$ and $(c - 1)(d - 1) \geq 0$. If

$$(a - 1)(b - 1) = (c - 1)(d - 1) = 1,$$

then each of $a - 1$, $b - 1$, $c - 1$ and $d - 1$ is 1, so that we have $(a, b, c, d) = (2, 2, 2, 2)$. Suppose

$$(a - 1)(b - 1) = 2 \text{ and } (c - 1)(d - 1) = 0.$$

Then $\{a - 1, b - 1\} = \{1, 2\}$ so that $cd = a + b = 5$. Hence $(a, b, c, d) = (2, 3, 1, 5)$, (2,3,5,1), (3,2,1,5) or (3,2,5,1). Similarly, if

$$(a - 1)(b - 1) = 0 \text{ and } (c - 1)(d - 1) = 2,$$

then $(a, b, c, d) = (1, 5, 2, 3)$, (5,1,2,3), (1,5,3,2) or (5,1,3,2). It is routine to check that all nine are indeed solutions.

Problem 2.

Does there exist a set of points in space having at least one but finitely many points on each plane?

182   *Hungarian Mathematical Olympiad (1964–1997)*

Solution:
We claim that the set $S = \{(t, t^3, t^5): t \text{ a real number}\}$ has the desired property. Consider any plane $ax + by + cz + d = 0$ where a, b and c are not all 0. Then the equation $at + bt^3 + ct^5 + d = 0$ has degree 1, 3 or 5, all of which are odd numbers. Hence it has at least one and at most five real solutions. It follows that this plane intersects S in at least one and at most five points.

Problem 3.
Let n be a positive integer. Among the $3n + 1$ members of a club, every two play exactly one of tennis, badminton and table tennis against each other. Each member plays each game against exactly n other members. Prove that there exist three members such that every two of them play a different game.

Solution:
Construct a graph with $3n + 1$ vertices representing the members. Two vertices are joined by a red edge if the members they represent play tennis against each other. Similarly, green edges are used for badminton and blue edges for table tennis. An arrow is defined as two edges having a common vertex, called its pivot. An arrow is said to be mismatched if its two edges have different colors. We now count the total number of mismatched arrows. At each vertex, there are n red edges, n green edges and n blue edges. Hence the number of mismatched arrows with this vertex as pivot is $3n^2$, and the grand total is $3n^2(3n + 1)$. Suppose none of the $\binom{3n+1}{3}$ triangles has three edges of different colors. Of the three arrows in this triangle, at most 2 are mismatched. Hence the total number of mismatched arrows is at most $2\binom{3n+1}{3} = n(3n + 1)(3n - 1) < 3n^2(3n + 1)$. This is a contradiction.

1988

Problem 1.
P is a point inside a convex quadrilateral $ABCD$ such that the areas of the triangles PAB, PBC, PCD and PDA are all equal. Prove that either AC or BD bisects the area of $ABCD$.

Solution:
Suppose P lies on AC. Then

$$[ABC] = [PAB] + [PBC] = [PCD] + [PDA] = [CDA]$$

and AC bisects $[ABCD]$. Suppose P does not lie on AC. Since we have $[PDA] = [PAB]$, the line AP passes through the mid-point of BD. Similarly, the line CP also passes through the midpoint of BD. Hence the common point P of AP and CP must be the midpoint of BD and thus lies on BD. It follows that BD bisects $[ABCD]$.

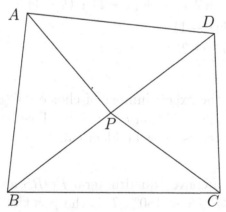

Problem 2.
From among the numbers $1, 2, \ldots, n$, we want to select triples (a, b, c) such that $a < b < c$ and, for two selected triples (a, b, c) and (a', b', c'), at most one of the equalities $a = a'$, $b = b'$ and $c = c'$ holds. What is the maximum number of such triples?

Solution:
Let us focus on (a, b, c) where b has a fixed value between 1 and n. Then the number of choices for a is $b - 1$ and the number of choices for c is $n - b$. Hence the number of choices for (a, b, c) is $\min\{b - 1, n - b\}$. If $n = 2k + 1$, we have

$$\sum_{b=2}^{n-1} \min\{b - 1, n - b\}$$
$$= 1 + 2 + \cdots + k + (k - 1) + \cdots + 1$$
$$= k^2$$
$$= \frac{(n-1)^2}{4}.$$

If $n = 2k$, we have instead

$$\sum_{b=2}^{n-1} \min\{b - 1, n - b\}$$
$$= 1 + 2 + \cdots + (k - 1) + (k - 1) + \cdots + 1$$
$$= k(k - 1)$$
$$= \frac{n(n-2)}{4}.$$

We can obtain the exact number of choice for (a, b, c) for each b by imposing the condition $a + b = c$. Then two triples that coincide in two terms must be identical.

Problem 3.
The vertices of a convex quadrilateral $PQRS$ are lattice points and $\angle SPQ + \angle PQR < 180°$. T is the point of intersection of PR and QS. Prove that there exists a lattice point other than P or Q which lies inside or on the boundary of triangle PQT.

Solution:
By symmetry, we may assume that $\angle PQR + \angle QRS \leq 180°$. Let S be closer to QR than P. Construct the parallelogram $PSRR'$. Since P, S and R are lattice points, so is R'. Note that R' is inside triangle PQR but cannot coincide with P or Q. If it is inside or on the boundary of triangle PQT, there is nothing further to prove. Assume now that R' is strictly inside triangle QRT. Then $PQR'S$ is a convex quadrilateral whose diagonals intersect at an interior point T'. We can repeat the construction above to obtain a new candidate S'. Since the number of lattice points inside triangle PQR is finite, the process must terminate in success at some point.

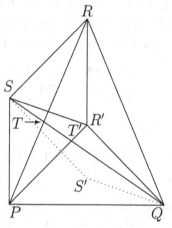

1989

Problem 1.

A circle is disjoint from a line m which is not horizontal. Construct a horizontal line such that the ratio of the lengths of the sections of this line within the circle and between m and the circle is maximum.

Solution:

In the special case where m is vertical, the desired horizontal line is clearly the one passing through the center of the circle. Henceforth, we assume that m intersects the vertical line n through the center of the circle at some point P. Let the desired horizontal line intersect the circle at A and B, intersect n at N and intersect m at M, with A between M and N. Minimizing

$$\frac{AM}{AB} = \frac{MN - AN}{2AN} = \frac{1}{2}\left(\frac{MN}{AN} - 1\right)$$

is the same as maximizing

$$\frac{AN}{MN} = \frac{AN}{PN} \cdot \frac{PN}{MN} = \frac{\tan PMN}{\tan PAN}.$$

Since $\angle MPN$ is fixed, the ratio is maximized when $\angle PAN$ is minimized, or equivalently, when $\angle APN$ is maximized. This occurs when PA is tangent to the circle. The other tangent from P will be PB, and AB is the desired horizontal line.

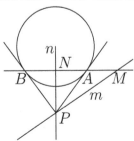

Problem 2.

For any positive integer m, denote by $s(m)$ the sum of its digits in base ten. For which positive integers m is it true that we have $s(mk) = s(m)$ for all integers k such that $1 \le k \le m$?

Solution:
Let the set of positive integers with the desired property be M. It is easy to verify that $1 \in M$, $2 \notin M$ and $9 \in M$. For any other $m \in M$, we have $m \ge 3$ so that we can choose $k = 3$. Then $s(m) = s(3m) \equiv 0 \pmod{3}$, so that m is a multiple of 3. This means that $3m$ is a multiple of 9. Hence we have $s(m) = s(3m) \equiv 0 \pmod{9}$, so that m is a multiple of 9 also. Thus $m \ge 10$, so that

$$m = a_0 10^n + a_1 10^{n-1} + \cdots + 10a_{n-1} + a_n$$

for some $n \ge 1$ and $a_0 \ge 1$. Choose $k = 10^n + 1 \le m$. Then

$$km = 10^n(m + a_0) + m - 10^n a_0.$$

Hence

$$
\begin{aligned}
s(m) &= s(km) \\
&= s(10^n(m + a_0)) + s(m - 10^n a_0) \\
&= s(m + a_0) + s(m) - a_0.
\end{aligned}
$$

It follows that $a_0 = s(m + a_0)$. The first digit of m is a_0. The first digit of $m + a_0$ cannot also be a_0 as otherwise we have $s(m + a_0) > a_0$. Hence it is 1, obtained from a chain of carrying-over. This implies that $a_0 = a_1 = \cdots = a_{n-1} = 9$. Since m is a multiple of 9, $a_n = 9$ also, and we must have $m = 10^{n+1} - 1$.

We now prove that $10^{n+1} - 1 \in M$. Let $k \leq m$. We may assume that k is not a multiple of 10 as otherwise the trailing 0s can be ignored. Then

$$
\begin{aligned}
s(km) &= s(10^{n+1}(k-1)) + s(10^{n+1} - k) \\
&= s(k-1) + 9n + 1 - s(k) \\
&= 9n \\
&= s(m).
\end{aligned}
$$

In conclusion, M consists of 1 along with $10^n - 1$ for $n \geq 1$.

Problem 3.
From an arbitrary point (x, y) in the coordinate plane, one is allowed to move to $(x, y+2x)$, $(x, y-2x)$, $(x+2y, y)$ or $(x, x-2y)$. However, one cannot reverse the immediately preceding move. Prove that starting from the point $(1, \sqrt{2})$, it is not possible to return there after any number of moves.

Solution:
Call a point (x, y) irrational if $\frac{y}{x}$ is irrational. In that case, $\frac{y}{x}$ is also irrational. In particular, $x \neq 0 \neq y$. For such a point (x, y), $(x, y \pm 2x)$ is also irrational since so is $\frac{y \pm 2x}{x} = \frac{y}{x} \pm 2$. Similarly, since $\frac{x \pm 2y}{y} = \frac{x}{y} \pm 2$, $(x \pm 2y, y)$ is also irrational. Starting from $(1, \sqrt{2})$ or any other irrational point for that matter, we can only move to other irrational points. If we never revisit any point, the desired conclusion follows. Henceforth we assume that there is a cycle joining distinct irrational points (x_0, y_0), (x_1, y_1), $\ldots$, (x_n, y_n) back to (x_0, y_0). Let (x_k, y_k) be a point on the cycle at maximum distance from $(0,0)$. By adjusting the indices if necessary, we may assume that $0 < k < n$. By symmetry, we may assume that $(x_{k-1}, y_{k-1}) = (x_k, y_k + 2x_k)$. Then $y_k^2 \geq (y_k + 2x_k)^2$. Since $(x_{k+1}, y_{k+1}) \neq (x_{k-1}, y_{k-1})$, we have two cases.
Case 1. $(x_{k+1}, y_{k+1}) = (x_k, y_k - 2x_k)$.
Then $y_k^2 \geq (y_k - 2x_k)^2$. Hence $0 \geq 8x_k^2$ so that $x_k = 0$. This is a contradiction.

Case 2. $(x_{k+1}, y_{k+1}) = (x_k \pm 2y_k, y_k)$.
Then $x_k^2 \geq (x_k \pm 2y_k)^2$. Hence

$$x_k^2 + y_k^2 \geq (x_k \pm 2y_k)^2 + (y_k + 2x_k)^2,$$

so that

$$0 \geq x_k^2 + y_k^2 + x_k y_k \pm x_k y_k.$$

If we take the minus sign, then $0 \geq x_k^2 + y_k^2$, which implies that $x_k = y_k = 0$. If we take the plus sign, then $0 \geq (x_k + y_k)^2$, which implies that $\frac{y_k}{x_k} = -1$. Both are contradictions.

1990

Problem 1.
Let p be any odd prime and n be any positive integer. Prove that at most one divisor d of pn^2 is such that $d+n^2$ is the square of an integer.

Solution:
Let d be a divisor of pn^2 such that $d+n^2 = m^2$ for some positive integer m. Let $pn^2 = kd$ and let g be the greatest common divisor of m and n. Then $\frac{m}{g}$ and $\frac{n}{g}$ are relatively prime. We have $pn^2 = k(m+n)(m-n)$ so that

$$p\left(\frac{n}{g}\right)^2 = k\left(\frac{m}{g}+\frac{n}{g}\right)\left(\frac{m}{g}-\frac{n}{g}\right).$$

Since p is prime, we must have

$$k = \left(\frac{n}{g}\right)^2,$$
$$p = \frac{m}{g}+\frac{n}{g},$$
$$1 = \frac{m}{g}-\frac{n}{g}.$$

Hence $\frac{n}{g} = \frac{p-1}{2}$ so that $g = \frac{2n}{p-1}$ is fixed. From $pn^2 = \left(\frac{n}{g}\right)^2 d$, $d = pg^2$ is also fixed.

Problem 2.
I is the incenter of triangle ABC. D, E and F are the excenters opposite A, B and C respectively. The bisector of $\angle BIC$ cuts BC at P. The bisector of $\angle CIA$ cuts CA at Q. The bisector of $\angle AIB$ cuts AB at R. Prove that DP, EQ and FR are concurrent.

Solution:

Let EQ and FR intersect at J, and let DJ and BC intersect at S. By the Sine Law, we have

$$\frac{DS}{\sin SCD} = \frac{SC}{\sin SDC},$$

$$\frac{QE}{\sin QCE} = \frac{QC}{\sin QEC},$$

$$\frac{DJ}{\sin QEC} = \frac{EJ}{\sin SDC}.$$

It follows that

$$DS \cdot QC \cdot EJ = SC \cdot EQ \cdot DJ,$$
$$EQ \cdot RA \cdot FJ = QA \cdot FR \cdot EJ,$$
$$FR \cdot SB \cdot DJ = RB \cdot DS \cdot FJ.$$

Hence we have $QC \cdot RA \cdot SB = SC \cdot QA \cdot RB$.

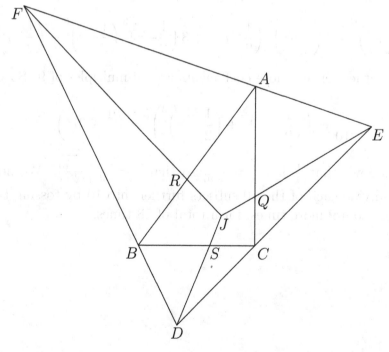

By the Angle Bisector Theorem, $\dfrac{IB}{IC} = \dfrac{PB}{PC}$, $\dfrac{IC}{IA} = \dfrac{QC}{QA}$ and $\dfrac{IA}{IB} = \dfrac{RA}{RB}$, so that $PB \cdot QC \cdot RA = PC \cdot QA \cdot RB$. Hence P and S coincide, and the desired result follows.

Problem 3.

A coin is tossed k times. Each time, the probability that it lands heads is p, where p is a real number between 0 and 1. Choose k and p for which the 2^k possible outcomes can be partitioned into 100 subsets such that the probability of the outcome being in any of the 100 subsets is the same.

Solution:

Let r be a real number between 0 and $\frac{1}{2}$. Let the coin lands heads with a probability of $\frac{1}{2} + r$, so that its probability of landing tails is $\frac{1}{2} - r$. We first solve the problem for 10 subsets. We choose $k = 9$. By the Binomial Theorem, $1 = ((\frac{1}{2} + r) + (\frac{1}{2} - r))^9$ is the sum of

$$\left(\frac{1}{2} + r\right)^9 + 3\left(\frac{1}{2} + r\right)^6 \left(\frac{1}{2} - r\right)^3 + 3\left(\frac{1}{2} + r\right)^3 \left(\frac{1}{2} - r\right)^6 + \left(\frac{1}{2} - r\right)^9$$

and other terms whose coefficients are all multiples of 9. So we set

$$\frac{1}{10} = \left(\left(\frac{1}{2} + r\right)^3 + \left(\frac{1}{2} - r\right)^3\right)^3 = \left(\frac{1}{4} + 3r^2\right)^3.$$

It follows that $\frac{1}{4} + 3r^2 = \frac{1}{\sqrt[3]{10}}$ so that $r = \sqrt{\frac{4 - \sqrt[3]{10}}{12\sqrt[3]{10}}}$. We now subdivide each of the 10 subsets further into 10 by tossing the same coin 9 more times, for a total of 18 times.

1991

Problem 1.

Prove that $\dfrac{(ab+c)^n - c}{(b+c)^n - c} \leq a^n$, where n is a positive integer and $a \geq 1$, $b \geq 1$ and $c > 0$ are real numbers.

Solution:

The desired inequality is equivalent to

$$a^n c - c \leq (ab + ac)^2 - (ab + c)^2.$$

After canceling the common factor $c(a-1) = (ab+ac) - (ab+c)$, we have

$$
\begin{aligned}
&a^{n-1} + a^{n-2} + \cdots + a + 1\\
\leq\ &(ab+ac)^{n-1} + (ab+ac)^{n-2}(ab+c)\\
&+ \cdots + (ab+ac)(ab+c)^{n-2} + (ab+c)^{n-1}.
\end{aligned}
$$

This inequality holds since for all k, $0 \leq k \leq n - 1$, we have

$$(ab+ac)^{n-k}(ab+c)^k \geq a^{n-k}a^{k-1} = a^{n-1} \geq a^{n-2} \geq \cdots \geq a \geq 1.$$

Problem 2.

A convex polyhedron has two triangular faces and three quadrilateral faces. Each vertex of one of the triangular faces is joined to the point of intersection of the diagonals of the opposite quadrilateral face. Prove that these three lines are concurrent.

Solution:

Let the triangular faces be ABC and DEF, connected by the edges AD, BE and CF. Let the diagonals of $BCEF$, $CAFD$ and $ABDE$ intersect at P, Q and R respectively. No two of the planes BCD, CAE and ABF are parallel. Hence they have a unique common point O.

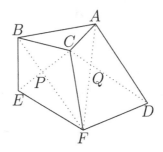

A clearly lies on the planes CAE and ABF. Since P lies on the lines CE and BE, it also lies on the planes CAE and ABF. Hence these two planes intersect along the line AP, so that O lies on AP. Similarly, O lies on BQ and CR, and we have the desired concurrency.

Problem 3.
Given are 998 red points in the plane, no three on a line. A set of blue points is chosen so that every triangle with all three vertices among the red points contains a blue point in its interior. What is the minimum size of a set of blue points which works regardless of the positions of the red points?

Solution:
More generally, we prove that for n red points, the minimum number of blue points is $2n - 5$. For $n = 998$, $2n - 5 = 1991$. We first establish necessity. Let the convex hull of the red points be a triangle $A_1A_2A_3$. Place A_4, A_5, ..., A_n on a circular arc from A_3 towards and above A_1A_2 as shown in the diagram below. Then no three of the red points are collinear. Joining each of A_4, A_5, ..., A_n to both A_1 and A_2 partitions $A_1A_2A_3$ into $2n-5$ non-overlapping triangles with red vertices, so that $2n - 5$ blue points are necessary.

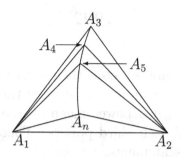

We now establish sufficiency. Consider the convex hull of n red points in general positions. Let there be i red points in the interior and b red points on the boundary. Then $i + b = n$. We claim that $2i + b - 2 = 2n - 2 - b$ blue points will be sufficient. The maximum value of this is $2n - 5$ if we take $b = 3$. For $1 \le k \le n$, let (x_k, y_k) be the coordinates of the red point A_k. Choose a coordinate system so that the y-coordinates are distinct, and relabel the red points if necessary so that $y_1 < y_2 < \cdots < y_n$. Since there are finitely many red points, we can choose a distance d which is less than any altitude of any triangle determined by three red points. We now add the blue points as follows. For each red point A_k in the interior of the convex hull, add the blue points $(x_k - d, y_k)$ and $(x_k + d, y_k)$. These points will be inside the convex hull. For each red points A_k on the boundary other than A_1 and A_n, add either $(x_k - d, y_k)$ or $(x_k + d, y_k)$, whichever is inside the convex hull. No blue points are added for A_1 and A_n. So the total number of blue points is indeed $2i + b - 2$. For any red triangle $A_p A_q A_r$ with $y_p < y_q < y_r$, either $(x_q - d, y_q)$ or $(x_q + d, y_q)$ will lie inside $A_p A_q A_r$. This justifies the claim.

1992

Problem 1.
Given n positive numbers, define their strange mean as the sum of the squares of the numbers divided by the sum of the numbers. Define their third power mean as the cube root of the arithmetic mean of their third powers. There are three mutually contradictory statements.

(1) The strange mean can never be smaller than the third power mean.

(2) The strange mean can never be larger than the third power mean.

(3) The strange mean may be larger or smaller than the third power mean.

Determine which of these statements is true for

(a) $n = 2$;

(b) $n = 3$.

Solution:

(a) Comparing $\frac{x^2+y^2}{x+y}$ with $\sqrt[3]{\frac{x^3+y^3}{2}}$ is the same as comparing $2(x^2+y^2)^3$ with $(x+y)^3(x^3+y^3)$. We have

$$
\begin{aligned}
& 2(x^2+y^2)^3 - (x+y)^3(x^3+y^3) \\
=\ & x^6 - 3x^5y + 3x^4y^2 - 2x^3y^3 + 3x^2y^4 - 3xy^5 + y^6 \\
=\ & (x^3 - y^3)^2 - 3x^4y(x-y) + 3xy^4(x-y) \\
=\ & (x-y)(x^3 - y^3)(x^2 + xy + y^2 - 3xy) \\
=\ & (x-y)^3(x^3 - y^3) \\
\geq\ & 0.
\end{aligned}
$$

Hence statement (1) is true for $n = 2$.

(b) On the one hand, $\frac{1^2+1^2+2^2}{1+1+2} = \frac{3}{2}$ while

$$\sqrt[3]{\frac{1^3 + 1^3 + 2^3}{3}} = \sqrt[3]{\frac{10}{3}}.$$

We have $(\frac{3}{2})^3 = \frac{27}{8} = \frac{81}{24} > \frac{80}{24} = \frac{10}{3}$. On the other hand, $\frac{2^2+2^2+3^2}{2+2+3} = \frac{17}{7}$ while

$$\sqrt[3]{\frac{2^3 + 2^3 + 3^3}{3}} = \sqrt[3]{\frac{43}{3}}.$$

We have $(\frac{17}{7})^3 = \frac{4913}{343} = \frac{14739}{1029} < \frac{14749}{1029} = \frac{43}{3}$. Hence statement (3) is true for $n = 3$.

Problem 2.

For an arbitrary positive integer k, let $f_1(k)$ be the square of the sum of the digits of k. For $n > 1$, let $f_n(k) = f_1(f_{n-1}(k))$. What is the value of $f_{1992}(2^{1991})$?

Solution:

Note that $f_1(k) \equiv k^2 \pmod 9$ for all positive integers k. Let $k = 2^{1991} \equiv 5 \pmod 9$. Then we have $f_1(k) \equiv 5^2 \equiv 7 \pmod 9$, $f_2(k) \equiv 7^2 \equiv 4 \pmod 9$, $f_3(k) \equiv 4^2 \equiv 7 \pmod 9$ as well as $f_4(k) \equiv 7^2 \equiv 4 \pmod 9$. Since $k < 2^{1992} = 8^{664} < 10^{664}$, we have

$$
\begin{aligned}
f_1(k) &< (664 \times 9)^2 < 6000^2 < 40000000, \\
f_2(k) &< f_1(39999999) < 70^2 < 5000, \\
f_3(k) &< f_1(4999) = 31^2 < 1000, \\
f_4(k) &< f_1(999) = 27^2.
\end{aligned}
$$

Now $f_4(k)$ is one of $7^2 = 49$, $16^2 = 256$ and $25^2 = 625$. In all cases, we have $f_5(k) = 13^2 = 169$. Hence $f_6(k) = 16^2 = 256$, so that $f_{2m+1}(k) = 169$ for all $m \geq 2$ and $f_{2m}(k) = 256$ for all $m \geq 3$. In particular, $f_{1992}(k) = 256$.

Problem 3.
Given a finite number of points in the plane, no three of which are collinear, prove that they can be painted in two colors so that there is no half-plane that contains exactly three given points of one color and no points of the other color.

Solution:
We will construct in stages a coloring scheme that works. Initially, consider the convex hull H of all the points. Paint those on the boundary of H red and those in the interior of H black. A half-plane which contains any black points must contain at least one red point. Suppose some half-plane contains exactly three red points. Let them be A_1, B_1 and C_1 that order along the boundary of H. Then none of the points is inside triangle $A_1B_1C_1$. Repaint B_1 black. If a half-plane contains exactly three black points, one of them must be B_1, but it must also contain either A_1 or C_1. Suppose some other half-plane contains exactly three red points. Let them be A_2, B_2 and C_2 in that order along the boundary of H. Then B_2 is distinct from A_1 and C_1. Repaint B_1 black. As before, no half-plane can contain exactly three black points. If there are still other half-planes containing exactly three red points, the middle point can be repainted. Since there are finitely many red points, the modification process must terminate at some coloring scheme that works.

1993

Problem 1.
Prove that if a and b are positive integers, then there are finitely many positive integers n for which both an^2+b and $a(n+1)^2+b$ are squares of integers.

Solution:
Let $an^2 + b = x^2$ and $a(n+1)^2 + b = y^2$ for positive integers x and y. Then $y^2 - x^2 - a = 2an$ so that

$$(y^2 - x^2 - a)^2 = 4a^2n^2 = 4a(x^2 - b).$$

Hence

$$
\begin{aligned}
4ab &= -(y^2 - x^2)^2 + 2a(y^2 - x^2) - a^2 + 4ax^2 \\
&= -(y - x)^2(y + x)^2 - a^2 + a(y + x)^2 + a(y - x)^2 \\
&= ((a - (y - x)^2)((y + x)^2 - a).
\end{aligned}
$$

Since $4ab$ is fixed, it has finitely many decompositions into two factors. Hence the number of choices for (x, y) is finite, and so is the number of choices for n.

Problem 2.
The sides of triangle ABC have different lengths. Its incircle touches the sides BC, CA and AB at points K, L and M, respectively. The line parallel to LM and passing through B cuts KL at point D. The line parallel to LM and passing through C cuts MK at point E. Prove that DE passes through the midpoint of LM.

Solution:
We have $\angle CEK = \angle LMK = \angle KLC$ by parallelism and tangency, respectively. Hence C, K, E and L are concyclic. Since $\angle ELM = \angle CEL = \angle CKL = \angle CLK = \angle EML$, we have $EM = EL$. Similarly, B, D, K and M are concyclic and $DM = DL$. Hence DE is the perpendicular bisector of LM, which passes through its midpoint.

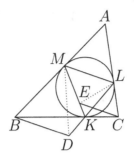

Problem 3.

Let $f(x) = x^{2n} + 2x^{2n-1} + 3x^{2n-2} + \cdots + 2nx + (2n+1)$, where n is a given positive integer. Find the minimum value of $f(x)$ for real numbers x.

Solution:

Note that

$$
\begin{aligned}
f(x) &= (x^{2n} + 2x^{2n-1} + x^{2n-2}) + 2(x^{2n-2} + 2x^{2n-3} + 2^{2n-4}) \\
&\quad + \cdots + (n-1)(x^4 + 2x^3 + x^2) + n(x^2 + 2x + 1) + n + 1 \\
&= (x+1)^2(x^{2n-2} + 2x^{2n-4} + \cdots + (n-1)x^2 + n) + n + 1 \\
&\geq n + 1.
\end{aligned}
$$

Since $f(-1) = (1-2)+(3-4)+\cdots+((2n-1)-2n)+2n+1 = n+1$, this is indeed the minimum value.

1994

Problem 1.
Let λ be the ratio of the sides of a parallelogram, with $\lambda > 1$. Determine in terms of λ the maximum possible measure of the acute angle formed by the diagonals of the parallelogram.

Solution:
Let $ABCD$ be the parallelogram with $\frac{AB}{AD} = \lambda > 1$. Let the diagonals intersect at E. Then $\angle AED < 90°$. Let $AE = a$, $BE = DE = b$ and $\angle AED = \theta$. We have

$$\lambda^2 = \frac{AB^2}{AD^2} = \frac{a^2 + b^2 + 2ab\cos\theta}{a^2 + b^2 - 2ab\cos\theta}.$$

Thus $\cos\theta = \dfrac{a^2 + b^2}{2ab} \cdot \dfrac{\lambda^2 - 1}{\lambda^2 + 1}$. Since the minimum value of $\frac{a^2+b^2}{2ab}$ is 1, attained when $a = b$, the maximum value of $\cos\theta$ is $\frac{\lambda^2-1}{\lambda^2+1}$. Hence the maximum measure of θ is $\arccos\frac{\lambda^2-1}{\lambda^2+1}$.

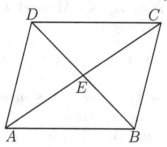

Problem 2.
Consider the diagonals of a convex n-gon.

(a) Prove that if any $n - 3$ of them are omitted, there are $n - 3$ remaining diagonals that do not intersect inside the polygon.

(b) Prove that one can always omit $n - 2$ diagonals such that among any $n - 3$ of the remaining diagonals, there are two which intersect inside the polygon.

Solution:

(a) More generally, we prove that if any k diagonals, $k \leq n-3$, are omitted, there are $n - 3$ remaining diagonals that do not intersect inside the polygon. We use induction on n. The result holds trivially for $n = 3$ and $n = 4$. Assume that it holds up to some $n \geq 4$. Consider an $(n + 1)$-gon $A_1 A_2 \ldots A_{n+1}$. If we do not omit any diagonals, the result certainly holds, as illustrated in the diagram below.

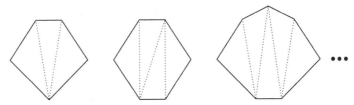

Henceforth, we assume that $k > 0$. We claim that there exists a j such that $A_{j-1} A_{j+1}$ is not omitted but at least one diagonal from A_j is. At most $k - 1$ diagonals are omitted from the n-gon

$$A_1 A_2 \ldots A_{j-1} A_{j+1} \ldots A_{n+1}.$$

Since $k \leq (n + 1) - 3$, $k - 1 \leq n - 3$. By the induction hypothesis, we can choose $n - 3$ non-intersecting diagonals. Choosing $A_{j-1} A_{j+1}$ as well in the $(n + 1)$-gon, we have $(n + 1) - 3$ diagonals as desired. Only the justification of the claim remains. If none of the diagonals $A_1 A_3$, $A_2 A_4$, $\ldots$, $A_n A_1$, $A_{n+1} A_2$ is omitted, such a j certainly exists. So we assume that at least one of them is omitted. Then there exists a j such that $A_{j-1} A_{j+1}$ is not omitted but $A_j A_{j+2}$ is. This is the desired j in the claim.

(b) From the convex n-gon $A_1A_2 \ldots A_n$, we omit the $n - 2$ diagonals A_nA_2, A_nA_3, $\ldots$, A_nA_{n-2} and $A_{n-1}A_1$. We are left with a convex $(n - 1)$-gon with none of its diagonals omitted. The number of diagonals that do not intersect inside the polygon is at most $(n - 1) - 3 = n - 4$. Hence among any $n-3$ of the remaining diagonals in $A_1A_2 \ldots A_n$, there are two which intersect inside the polygon.

Problem 3.

For $1 \leq k \leq n$, the set H_k consists of k pairwise disjoint intervals of the real line. Prove that among the intervals that form these sets, one can find $\lfloor \frac{n+1}{2} \rfloor$ pairwise disjoint intervals, each of which belongs to a different set.

Solution:

We use induction on n. The result certainly holds for $n = 1$ and $n = 2$ since $\lfloor \frac{n+1}{2} \rfloor = 1$. Assume that it holds up to n for some $n \geq 2$. Consider the case $n + 1$. Choose among all $n + 1$ sets the interval I whose right endpoint is minimum. If there are two or more choices, pick one arbitrarily. Assume that I belongs to a specific H_m. Delete the set H_m. From every set other than H_m, delete the interval whose right endpoint is minimum. Then I is disjoint from every remaining interval. The modified sets H'_1, H'_2, $\ldots$, H'_{m-1}, H'_{m+1}, $\ldots$, H'_{n+1} contain respectively, 0, 1, $\ldots$, $m - 2$, m, $\ldots$, n intervals. From each H'_k, $m + 1 \leq k \leq n + 1$, delete an arbitrary interval. Then the numbers of intervals in the re-modified sets H''_k become $m - 1$, $\ldots$, $n - 1$. Apply the induction hypotheses to the $n - 1$ sets H'_2, $\ldots$, H'_{m-1}, H''_{m+1}, $\ldots$, H''_{n+1}, we can choose $\lfloor \frac{(n-1)+1}{2} \rfloor$ pairwise disjoint intervals, each belonging to a different set. They are all disjoint from I, so that the total number of intervals chosen is $\lfloor \frac{n}{2} \rfloor + 1 = \lfloor \frac{(n+1)+1}{2} \rfloor$, completing the inductive argument.

1995

Problem 1.

A lattice rectangle with sides parallel to the coordinate axes is divided into lattice triangles, each of area $\frac{1}{2}$. Prove that the number of right triangles among them is at least twice the length of the shorter side of rectangle.

Solution:

By Pick's Formula, a lattice triangle of area $\frac{1}{2}$ contains exactly 3 boundary lattice points and 0 interior lattice points. If it has a right angle, then it must be a half-square. Two half-squares can be put together to form a parallelogram with sides 1 and $\sqrt{2}$. These are called basic parallelograms.

Suppose that two of the triangles in the dissection of the overall rectangle R form a convex quadrilateral. Then each diagonal divides it into two triangles of area $\frac{1}{2}$. It follows that the diagonals bisect each other, so that the quadrilateral is a parallelogram. If it is a square or a basic parallelogram, we leave it alone. Otherwise, the two diagonals will be of different length, the longer of which has length exceeding $\sqrt{5}$. If the longer diagonal is the dividing line between the two triangles, we replace it with the shorter one. The replacement reduces the total perimeter of all the triangles, without creating new half-squares. Since the total perimeter cannot decrease indefinitely, the process must terminate at some point.

Now the dissection consists only of half-squares and basic parallelograms. Consider a unit segment on the boundary of R. If it is a side of a half-square, we stop. It it is a side of a basic parallelogram, we move over to the opposite side which is another unit segment, and continue from there. We either terminate in a half-square or reach the opposite side of R.

In the latter case, we have a chain of basic parallelograms. Note that it is not possible to have two chains, each connecting a different pair of opposite sides of R. This is because two such chains can only cross each other at a square, which is not a basic parallelogram. So, starting from each unit segment on one pair of opposite sides of R, we always end up in a half-square. These half-squares are distinct as each can only be the final stop for two unit segments on adjacent sides of R. The desired conclusion now follows.

Problem 2.
Each of the n variables of a polynomial is substituted with 1 or -1. If the number of -1s is even, the value of the polynomial is positive. If it is odd, the value is negative. Prove that the polynomial has a term in which the sum of the exponents of the variables is at least n.

Solution:
We may assume that the exponent of each variable in each term is at most 1, since $1^2 = (-1)^2$. A square-free polynomial in n variables is the sum of $Ax_1x_2 \cdots x_n + C$ plus terms of the form $Bx_{i_1}x_{i_2} \cdots x_{i_k}$, where $1 \leq k \leq n-1$. There are 2^n different ways of substituting 1 or -1 for the variables. In 2^{n-1} ways, the number of -1s used is even. Let their sum be denoted by S_n. In the other 2^{n-1} ways, the number of -1s used is odd. Let their sum be denoted by T_n. Consider a typical term $x_1x_2 \cdots x_k$ which does not feature in S_j or T_j for $j < k$. Clearly, it contributes 2^{k-1} to S_k and -2^{k-1} to T_k. The same two contributions are made to S_{k+1} so that the net contribution is 0. The same is true for its contribution to T_{k+1}, and its contribution to S_j and T_j are both 0 for all $j > k$. It follows that $S_n = 2^{n-1}(A + C)$ and $T_n = 2^{n-1}(-A + C)$. From $A + C > 0$ and $-A + C < 0$, we have $A > 0$, from which the desired conclusion follows.

Problem 3.
No three of the points A, B, C and D are collinear. Let E and F denote the points of intersection of the lines AB and CD, and of the lines BC and DA, respectively. Circles are drawn with the segments AC, BD and EF as diameters. Show that either the three circles have a common point or they are pairwise disjoint.

Solution:
We first prove an auxiliary result. Let P, Q and R be the respective midpoints of DF, FC and CD. Then QR intersects AC at its midpoint K, RP intersects BD at its midpoint L and PQ intersects EF at its midpoint M. By similar triangles, we have

$$\frac{QK}{KR} \cdot \frac{RL}{LP} \cdot \frac{PM}{MQ} = \frac{FA}{AD} \cdot \frac{CB}{CF} \cdot \frac{DE}{EC}.$$

Applying Menelaus's Theorem to triangle CDF with respect to the transversal ABE, the right side of the above equation is equal to 1. Hence so does the right side. By the converse of Menelaus' Theorem applied to triangle PQR, K, L and M are collinear.

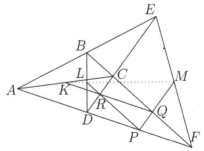

Let the feet of the altitudes of triangle ADE from A, D and E be U, V and W respectively. Then U lies on α since AC is its diameter. Similarly, V lies on β and W lies on γ. The altitudes AU, DV EW are concurrent at the orthocenter H, and we have $HA \cdot HU = HD \cdot HV = HE \cdot HW$. Hence H has equal power with respect to each of α, β and γ.

Since the centers of these circles are collinear, their pairwise radical axes are parallel. Since they all pass through H, they must coincide. If the circles are pairwise disjoint, there is nothing further to prove. If any two of them have a common point, it must lie on the common radical axis and therefore also on the third circle.

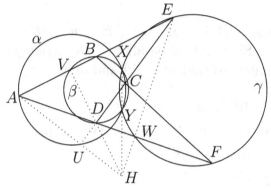

1996

Problem 1.
In the quadrilateral $ABCD$, AC is perpendicular to BD and AB is parallel to DC. Prove that $BC \cdot DA \geq AB \cdot CD$.

Solution:
Let AC intersect BD at E. Let $AE = w$, $BE = x$, $CE = y$ and $DE = z$. Since AB is parallel to DC, triangles ABE and CDE are similar, so that $w \geq y$ if and only if $x \geq z$. Now

$$
\begin{aligned}
& BC^2 DA^2 - AB^2 CD^2 \\
={}& (x^2 + y^2)(z^2 + w^2) - (w^2 + x^2)(y^2 + z^2) \\
={}& (w^2 - y^2)(x^2 - z^2) \\
\geq{}& 0.
\end{aligned}
$$

It follows that $BC \cdot DA \geq AB \cdot CD$.

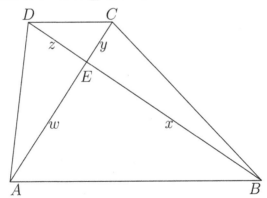

Problem 2.
The same numbers of delegates from countries A and B attend a conference. Some pairs of them already know each other. Prove that there exists a non-empty set of delegates from country A such that either every delegate from country B has an even number of acquaintances among them, or every delegate from country B has an odd number of acquaintances among them.

Solution:
Let the common number of delegates be n. Each of the $2^n - 1$ non-empty subsets of delegates from country A is associated with a binary n-tuple $(b_1, b_2, \ldots, b_n)$ where for $1 \leq i \leq n$, $b_i = 0$ if the ith delegate from country B has an even number of acquaintances among the members of the subset, and $b_i = 1$ otherwise. There are 2^n different n-tuples. If any of them is $(0,0,\ldots,0)$ or $(1,1,\ldots,1)$, there is nothing further to be proved. Assume that neither exists. By the Pigeonhole Principle, two of the n-tuples must be identical. Let them be associated with the subsets P and Q. Consider the subset which consists of delegates from country A who belong to exactly one of P and Q. Then this subset must be associated with $(0,0,\ldots,0)$, which is a contradiction.

Problem 3.
For integers $n \geq 3$ and $k \geq 0$, mark some of the diagonals of a convex n-gon. We wish to choose a polygonal line consisting of $2k + 1$ marked diagonals and not intersecting itself.

(a) Prove that this is always possible if $2kn + 1$ diagonals are marked.

(b) Prove that this may not be possible if kn diagonals are marked.

Solution:
We define the length of a diagonal as ℓ if the shorter way from one endpoint of the diagonal along the perimeter of the n-gon to the other endpoint passes over ℓ sides of the n-gon.

(a) We use mathematical induction on k to prove that if any $2kn + 1$ diagonals of a convex n-gon are marked, we can choose an *open* polygonal line which consists of $2k + 1$ of the marked diagonals and does not intersecting itself. The result holds trivially for $k = 0$ since any one marked diagonal will constitute the desired polygonal line. Suppose the result holds for some $k \geq 0$. We now mark $2(k + 1)n + 1$ diagonals of the n-gon.

Consider the vertices one at a time. If it is the endpoint of only one marked diagonal, unmark that diagonal. If it is the endpoint of at least two marked diagonals, unmark the two diagonals whose other endpoints are closest to the vertex under consideration, one on each side along the perimeter of the n-gon. Note that some diagonals may already been unmarked when subsequent vertices are considered. When this process is over, we have unmarked at most $2n$ diagonals, so that at least $2kn + 1$ marked diagonals remain.

By the induction hypothesis, we can choose an open polygonal line $A_0 A_1 \ldots A_{2k} A_{2k+1}$ with $A_0 \neq A_{2k+1}$. The entire polygonal line lies on one side of $A_0 A_1$. Now two diagonals with A_0 as an endpoint have been unmarked, because $A_0 A_1$ still survives. One of them is in the opposite side to the polygonal line of the line $A_0 A_1$. It can be retrieved and added to the polygonal line. A similar extension occurs at A_{2k+1}, and we have an open polygonal line with the desired properties having length $2k + 1 + 2 = 2(k + 1) + 1$.

(b) Let $A_1 A_2 \ldots A_r$ be the longest polygonal line not intersecting itself that can be chosen from the marked diagonals. Then the entire polygonal line lies on one side of $A_1 A_2$, and also on one side of $A_{r-1} A_r$. We consider two cases.

Case 1. $n = 2m + 1$ for some integer m.

We mark all diagonals of lengths $m - k + 1$, $m - k + 2$, ..., m, a total of

$$n(m - (m - k + 1) + 1) = kn$$

diagonals. There are $m - k$ vertices of the n-gon between A_1 and A_2, and another $m-k$ between A_{r-1} and A_r. Hence $r \le n - 2(m - k) = 2k + 1$, so that there are at most $2k$ diagonals on the polygonal line.

Case 2. $n = 2m$ for some integer m.

We mark all n diagonals of each of the lengths $m - k$, $m - k + 1$, ..., $m - 1$, a total of

$$n((m - 1) - (m - k) + 1) = kn$$

diagonals. Now the vertices A_1, A_2, A_{r-1} and A_r divide the perimeter of the n-gon into four parts. The parts between A_1 and A_2 contains $m - k - 1$ other vertices, as does the part between A_{r-1} and A_r. However, this only shows that $r \le n - 2(m - k - 1) = 2k + 2$, and we need a refinement. Consider the other two parts of the perimeter. If both endpoints of a diagonal $A_h A_{h+1}$ along the polygonal line are in the same part, then the entire polygonal line lies on one side of it. Since $m - k \ge 2$, we have

$$
\begin{aligned}
r &\le n - 3(m - k - 1) \\
&= 2k + 1 - (m - k - 2) \\
&\le 2k + 1.
\end{aligned}
$$

On the other hand, if the vertices of the polygonal line lie alternately in the two parts, then the lengths of the diagonals first increase and then decrease, and each possible length appears at most twice. Since there are

$$(m - 1) - (m - k) + 1 = k$$

possible lengths, the number of diagonals is at most $2k$.

1997

Problem 1.

Let p be an odd prime number. Consider points in the coordinate plane both coordinates of which are numbers in the set $\{0, 1, 2, \ldots, p-1\}$. Prove that it is possible to choose p of these points such that no three are collinear.

Solution:

For $0 \le i \le p-1$, let $x_i = i$ and $y_i \equiv x_i^2 \pmod{p}$, reduced so that $0 \le y_i \le p-1$. Consider any three of these p points (x_i, y_i), (x_j, y_j) and (x_k, y_k). In arithmetic modulo p, the area of the triangle they determine is given by

$$x_i y_j + x_j y_k + x_k y_i - x_i y_k - x_j y_i - x_k y_j$$
$$\equiv x_i x_j^2 + x_j x_k^2 + x_k x_i^2 - x_i x_k^2 - x_j x_i^2 - x_k x_j^2$$
$$= (x_i - x_j)(x_j - x_k)(x_k - x_i)$$
$$\neq 0.$$

It follows that they are not collinear.

Problem 2.

The incircle of triangle ABC touches the sides at D, E and F. Prove that its circumcenter and incenter are collinear with the orthocenter of triangle DEF.

Solution:

In triangle ABC, let O be the center and R be the radius of the circumcircle, and let I be the center and r be the radius of the incircle. Let AI, BI and CI intersect the circumcircle again at L, M and N respectively. Let MN intersect AI at X, NL intersect BI at Y and LM intersect CI at Z. Note that O is also the circumcenter of triangle LMN. Now L is the midpoint of the arc BC, M is the midpoint of the arc CA and N is the midpoint of the arc AB. It follows that we have $\angle CNL = \angle CAL = \frac{1}{2}\angle CAB$, $\angle CNM = \angle CBM = \frac{1}{2}\angle ABC$ and $ALN = \angle ACN = \frac{1}{2}\angle BCA$.

Now

$$\angle NXL = 180° - (\angle CNL + \angle CNM + \angle ACN) = 90°.$$

Hence MN is perpendicular to LI. Similarly, NL is perpendicular to MI and LM is perpendicular to NI, so that I is the orthocenter of LMN. It follows that the line ℓ determined by O and I is its Euler line.

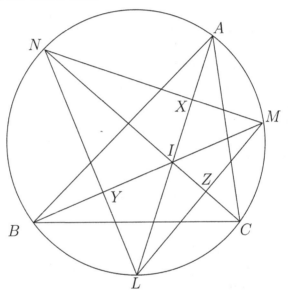

Let D, E and F be the points of tangency with BC, CA and AB respectively. Then I is also the circumcenter of triangle DEF. The center J of homothety of the two circles lies on ℓ. The ratio of homothety is $\frac{r}{R}$. Note that both OL and ID are perpendicular to BC, so that they are parallel. Moreover, $OL = R$ and $ID = r$. Hence L is indeed the homothetic image of D. Similarly, M is the homothetic image of E and N is the homothetic image of F. It follows that DEF and LMN are homothetic about J. Since ℓ is the Euler line of LMN, it is also the Euler line of DEF. Hence its orthocenter lies on ℓ, which is the desired result.

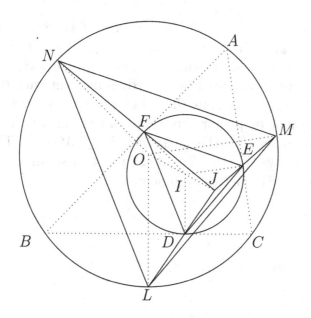

Problem 3.

Prove that the edges of a planar graph can be painted in three colors so that each cycle contains at least two edges of different colors.

Solution:

Let the colors be red, blue and green. We use induction on the number of edges of the planar graph. The basis is trivial. Remove an arbitrary edge xy. Then the remaining edges can be properly 3-colored. It follows that there must be a red path $R = (r_0r_1, r_1r_2, \ldots, r_{i-1}r_i)$ connecting $x = r_0$ and $y = r_i$, as otherwise we can just paint xy red. This path is unique as otherwise we will have a red cycle. Similarly, there is a unique blue path $B = (b_0b_1, b_1b_2, \ldots, b_{j-1}b_j)$ and a unique green path $G = (g_0g_1, g_1g_2, \ldots, g_{k-1}g_k)$ connecting x and y. If r_t and r_{t+1} are not connected by a blue path, we can repaint it blue and paint xy red. Hence every two adjacent vertices on R are connected by a blue path. Taking B into consideration, we will have blue cycles unless $\{r_0, r_1, \ldots, r_i\}$ is a subset of $\{b_0, b_1, \ldots, b_j\}$.

By symmetry, $\{b_0, b_1, \ldots, b_j\}$ is a subset of $\{g_0, g_1, \ldots, g_k\}$ and $\{g_0, g_1, \ldots, g_k\}$ is a subset of $\{r_0, r_1, \ldots, r_i\}$. Thus the three sets are just permutations of some $\{x, v_1, v_2, \ldots, v_n, y\}$. Here we have $n + 2$ vertices and at least $3n + 4$ edges incident with them, including the edge xy. Thus the subgraph induced by this set of vertices cannot be planar, since a planar graph with V vertices and E edges satisfies the inequality $E \leq 3V - 6$. This contradiction shows that R, B and G cannot exist simultaneously, completing the inductive argument.

Part IV: Appendix

From September 1968 to May 1984, KÖMAL published some "Problems beyond Competition". The statements are given below. The solutions are left to the reader.

1. There are three stops on every bus line. It is possible to transfer between any two bus lines, but only at one stop. It is possible to go from each stop to any other stop, making at most one transfer. How many bus lines are there?

2. Prove that a bound K can be given such that if we take those positive integers which are less than an arbitrary number N and which do not contain the digit 7, then the sum of their reciprocals is less than K.

3. Let $p(x)$ be a polynomial whose coefficients are integers. Prove that $6(x^2 + 1)^2 + 5(x^2 + 1)p(x) - 21(p(x))^2$ has no integer roots.

4. Let α be a root of the equation $x^2 + px + q = 0$ and β be a root of the equation $x^2 + rx + s = 0$. If $\alpha = \beta k$, prove that $(q - k^2 s)^2 + k(p - kr)(kps - qr) = 0$.

5. On each side of a hexagon, construct an equilateral triangle outside the hexagon. The six new vertices determine a new hexagon. If the midpoints of the sides of this hexagon form a regular hexagon, prove that the original hexagon has central symmetry.

6. Construct a quadrilateral $ABCD$ with AD parallel to BC, given its perimeter, the length of the diagonal AC and the measures of $\angle BAD$ and $\angle CDA$.

7. The last k digits of all integral powers of a k-digit integer are identical to the digits of the integer itself. Determine all such integers.

8. There are 3 lamps operated by n switches. Each switch may be in one of 3 different positions. In every configuration of the switches, at least 1 lamps is lit. If the position of every switch is arbitrarily changed at the same time, then a different lamp will be lit. Prove that there is a single switch whose position controls which lamp is lit.

9. Construct a cyclic quadrilateral given the lengths of its four sides.

10. Let C be the center of a circle γ.

    (a) The inversive image of a line ℓ not passing through C with respect to γ is a circle passing through C. Prove that its center is the reflectional image of C across ℓ.

    (b) The inversive image of a circle ω not passing through C with respect to γ is also a circle not passing through C. Prove that its center is the inversive image of C with respect to ω.

11. The point $P_2(\frac{1}{2}, \frac{\sqrt{3}}{2}, 0)$ is at a distance 1 from both $P_0(0,0,0)$ and $P_1(1,0,0)$. Determine the coordinates of the point P_3 which is at a distance 1 from all three points P_1, P_2 and P_3.

12. In an attempt to solve the equation $t^3 + 2t^2 + 5t + 2 = 0$, we write $t = \frac{t^3 + 2t^2 + 2}{5}$. Using t_0 as an initial approximation, we define $t_{n+1} = \frac{t_n^3 + 2t_n^2 + 2}{5}$ for $n \geq 0$.

    (a) If we choose $t_0 = -3, -2, -1, -\frac{1}{2}, 0, 1$ or 2, is it true that from some point on, the sequence $\{t_n\}$ always increases or always decreases?

    (b) Is it true that the equation has a root for which an arbitrarily good approximation may be obtained this way?

13. In an attempt to construct a regular 17-gon $P_1 P_2 \ldots P_{17}$, we start with its circumcircle with center O. Let P_1 be an arbitrary point on the circle, OC be a radius perpendicular to OP_1 and D be the point on OC such that $CD = 3OD$. Let E be the point on OP_1 such that $\angle P_1 DE = 3\angle EDO$, and let F be the point on the extension of $P_1 O$ such that $\angle EDF = 45°$. Let the circle with diameter $P_1 F$ intersect OC at G, and let the circle with center E and radius EG intersect OP_1 at H and the extension of $P_1 O$ at K. Let the lines through H and K perpendicular to OC intersect the same half of the circumcircle, say the one containing C, at P_4 and P_6 respectively. Let P_5 be the midpoint of the minor arc $P_4 P_6$. Then $P_4 P_5$ and $P_5 P_6$ are sides of the 17-gon. The remaining vertices are now easy to obtain. Is this construction precise, or is it only an approximation?

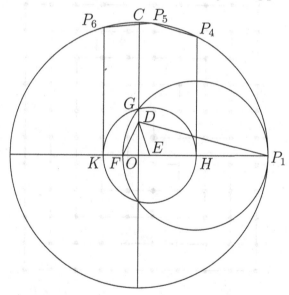

14. The point P is inside $\angle XOY$. The line ℓ passing through O is not inside $\angle XOY$. Let Q be an arbitrary point on the segment OP. Construct a line through Q parallel to ℓ, intersecting OX and OY at E and F respectively.

(a) Determine that position of Q for which the area of triangle OEF is maximal.

(b) Determine the position of ℓ for which th maximal area of OEF is minimal.

15. A grasshopper visits all the lattice points in the plane, moving towards $(0,0)$ along the negative x-axis. Upon reaching $(0,0)$, it begins to zig-zag in the pattern shown in the diagram. The point $(0,0)$ is labeled 0, the point $(0,1)$ is labeled 1, the point $(1,1)$ is labeled 2, and so on.

(a) What is the label of the point (n, n)?

(b) Give bounds for the Euclidean distance from $(0,0)$ to the point with label n?

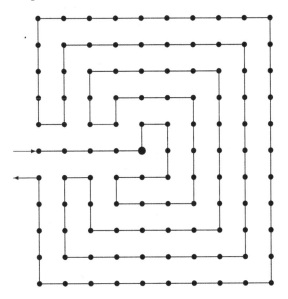

16. For how many integer values of n will

$$4n^4 - 12n^3 + 17n^2 - 6n - 14$$

be the square of an integer?

17. Two circles intersect at the points A and P. A variable line through P intersects the circles again at B and C respectively. What are the loci of the incenter, circumcenter, the centroid and the orthocenter of triangle ABC?

18. The four vertices of a convex quadrilateral are to be connected by a network of line segments. How should the network be constructed so that the total length of the segments in the network is minimum?

19. A is the union of n disjoint open intervals and B is the union of m disjoint open intervals on the real line. At most how many disjoint open intervals can $A \cap B$ have?

20. Prove that $\displaystyle\prod_{i=1}^{n}\left(i + \frac{1}{n(n!)}\right) < n! + 1$ for all positive integers n.

21. Determine all positive integers whose digits can be partitioned at a suitably chosen point into two parts such that the product of the numbers formed from the digits in each part is equal to half of the original number.

22. Construct a triangle DEF congruent to a given triangle ABC such that the lines AE, BF and CD are parallel to one another, and the two triangles are

    (a) in the same orientation;

    (b) in opposite orientations.

23. Let $a_1 = 1$ and $a_{n+1} = n^{a_n}$ for $n \geq 1$. Determine the last digit of a_n in terms of n.

24. A knight wishes to visit in consecutive moves every square of a $4 \times k$ chessboard exactly once.

(a) Prove that this is impossible if $k = 4$.

(b) Prove that this is possible for $k = 3$ and for $k \geq 5$ only if the knight, in the first half of the tour, visits all squares on the first and fourth rows of one color and all squares on the second and third rows of the opposite color.

25. Here is a method devised by the great Hungarian mathematician *János Bolyai* for trisecting an acute angle. Place its vertex at the origin O, one arm along the x-axis OX and the other arm through a point P in the first quadrant on the curve $xy = 1$. Draw a circle with center P and radius $2OP$, cutting the curve at two points. Let the point closer to OX be Q, and draw the ray PR from P in the direction of the positive x-axis. Then $\angle QPR = \frac{1}{3}\angle POX$. Prove that this construction is exact.

26. Given a point P, two lines ℓ and m, and two acute angles α and β. Construct a circle passing through P and intersecting ℓ and m at A and B respectively, such that the angle formed by ℓ and the tangent to the circle at A is α, and the angle formed by m and the tangent to the circle at B is β.

27. Let $n \geq 2$ be an integer. Consider the system of equations

$$
\begin{aligned}
x_2 &= ax_1^2 + bx_1 + c, \\
x_3 &= ax_2^2 + bx_2 + c, \\
\cdots &= \cdots, \\
x_n &= ax_{n-1}^2 + bx_{n-1} + c, \\
x_1 &= ax_n^2 + bx_n + c.
\end{aligned}
$$

Prove that the real coefficients a, b and c can be chosen so that the system has at least three solutions.

28. A circle is divided into two unequal segments by a chord. A circle is inscribed into the larger segment. For which choice of the chord does the area of the part of the larger segment outside the inscribed circle attains its maximum value?

29. Let C be the center of a circle γ.

   (a) Let ℓ be a line not passing through C. Its inversive image with respective to γ is a circle ℓ' passing through C. Let P and Q be reflectional images of each other across ℓ and let P' and Q' be their respective inversive images with respect to γ. Prove that P' and Q' are inversive images of each other with respect to ℓ'.

   (b) Let ω be a circle not passing through C. Its inversive image with respective to γ is another circle ω' not passing through C. Let P and Q be inversive images of each other with respect to ω' and let P' and Q' be their respective inversive images with respect to γ. Prove that P' and Q' are inversive images of each other with respect to ω'.

   (c) Use part (a) to deduce part (a) of Problem 12.

   (d) Use part (b) to deduce part (b) of Problem 12.

30. $ABCD$ is a convex quadrilateral. A, B, C and D are joined to the midpoints of BC, CD, DA and AB respectively. Prove that the area of the convex quadrilateral determined by these four lines is at least $\frac{1}{6}$ and at most $\frac{1}{5}$ the area of $ABCD$.

31. Either compute $\lim\limits_{x \to 0} x \left\lfloor \dfrac{1}{x} \right\rfloor$ or prove that it does not exist.

32. Five different numbers are drawn at random from the first 90 positive integers.

(a) Find the probability that the sum of the smallest two numbers is the third smallest number?

(b) What is the most likely value of the third smallest number if it is the sum of the smallest two numbers?

33. Determine all positive integers m and n less than 100 such that $(m + n)(100m + n) = m^3 + n^3$.

34. A line ℓ intersects the lines containing the sides BC, CA and AB of triangle ABC at D_1, E_1 and F_1 respectively. Similarly, a perpendicular to ℓ generates the points D_2, E_2 and F_2. Prove that the midpoints of D_1D_2, E_1E_2 and F_1F_2 are collinear.

35. We wish to express positive integers as a sum of three distinct positive integers. Two ways involving the same three numbers are not considered different. There are $3^3 = 27$ different ways of expressing 21, $3^5 = 243$ different ways of expressing 57 and $3^7 = 2187$ ways of expressing 165. Is there some general pattern?

36. Let p be a prime and k be a positive integer. Prove that if $n!$ is divisible by p^k, then it is also divisible by $(p!)^k$.

37. A circle is tangent to two lines which are perpendicular to each other. A third line these two lines at P and Q and the circle at R and S. Given only these four points, reconstruct the whole figure.

38. Is there a convex polyhedron such that a square and four equilateral triangles meet at each of its vertices?

39. Is it possible to label each of the ten dots in the diagram with a different number chosen from 1, 2, 3, 4, 5, 6, 7, 8, 9 and 10 such that the sum of the labels of the four dots on each of the five lines is the same?

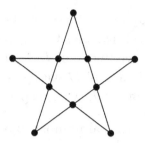

40. (a) Consider the infinite sequence in which the first term is 1 and all subsequent terms are 0s. Starting from the first term, the subsequences of lengths 1, 2, 3, ... are (1), (1,0), (1,0,0), ... If we string these subsequences together, we obtain another infinite sequence (1,1,0,1,0,0,...). Prove that this sequence is not periodic.

(b) Consider the periodic infinite sequence which keeps repeating the cycle (1,2,3). Starting from the first term, the subsequences of even lengths 2, 4, ... are (1,2), (1,2,3,1), ... If we string these subsequences together, we obtain another infinite sequence

$$(1, 2, 1, 2, 3, 1, \ldots).$$

Prove that this sequence is not periodic.

(c) If we add the corresponding terms of the infinite sequences obtained in (a) and (b), we obtain the infinite sequence (2,3,1,3,3,1,...). Prove that this sequence is not periodic.

41. Let a, b, c, p, q and r be real numbers such that $a > 0$, $p > 0$, $ab - c^2 > 0$ and $pq - r^2 > 0$. Prove that

$$\frac{8}{(a+p)(b+q) - (c+r)^2} \leq \frac{1}{ab - c^2} + \frac{1}{pq - r^2}.$$

42. The circles ω_1, ω_2 and ω_3 have equal radii and their centers form an equilateral triangle. Construct a point P such that if Q is the inversive image of P with respect to ω_1 and R is the inversive image of Q with respect to ω_2, then P is the inversive image of R with respect to ω_3.

43. Let x be a non-zero real number and n be a positive integer. Express $x^n + \frac{1}{x^n}$ in terms of $x + \frac{1}{x}$.

44. What is the behavior of the roots of the quadratic equation $ax^2 + bx + c = 0$ if b and c are constants while a approaches 0?

45. The circles ω and γ intersect each other at A and B. The radius of ω is greater than that of γ. A line is tangent to ω at T and γ at U. A line through A parallel to TU intersects ω at C and γ at D. E is the point on the extension of CD such that $\frac{EC}{ED} = \frac{AC}{AD}$. The circumcircles of triangles ABE and ATU intersect again at F. Prove that AF is perpendicular to TU.

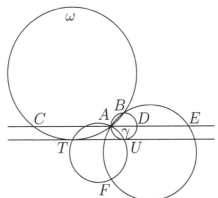

46. Starting from a triangle determined by three vertices of a regular n-gon, we construct a sequence of triangles, each one being determined by the feet of the three altitudes of the preceding triangle. For $n = 5$ or 7, some triangle in this sequence is similar to the starting one. For what other values of n is this result true?

47. Let p be a prime number. The numbers $0, 1, \ldots, p - 1$ are divided arbitrarily into two classes. Prove that there is at most one number a among them such that neither a nor $p + a$ can be expressed as the sum of one number from each class.

48. Prove that in a table-tennis tournament in which each participant plays once against every other participant, all the participants can be arranged in a line such that each one after the first has beaten the participant immediately preceding her.

49. In triangle ABC, P is the reflectional image of B across the bisector of $\angle C$ and Q is the reflectional image of C across the bisector of $\angle B$. Prove that PQ is perpendicular to the line joining the circumcenter and the incenter of ABC.

50. Given two perpendicular lines and two points not on either of them, construct an ellipse that passes through both points and tangent to both lines, each at one endpoint of an axis of the ellipse.

51. Prove that among five lattice points in the plane, there exist two such that the line segment joining them passes through at least one lattice point.

52. Prove that $\displaystyle\sum_{k=n}^{2n} \binom{m}{k}\binom{k}{2n-k} 2^{2k-2n} = \binom{2m}{2n}$ for positive integers $m > k > n$.

53. The diagram shows an infinite table with three rows. The numbers in the first row decrease to 14 and then increase thereafter. The numbers in the second row are increasing while the numbers in the third row are decreasing. The sum of the numbers in each column is 8, and the sum of the squares of the numbers in each column is the cube of an integer. Show how the table can be extended in both directions.

$\cdots$	86	26	14	50	134	$\cdots$
$\cdots$	-130	-18	-2	14	126	$\cdots$
$\cdots$	52	0	-4	-56	-252	$\cdots$

54. Three points are chosen at random among 180 evenly distributed points on a circle. What is the probability that in the triangle they determine, each angle is at least 30°?

55. The diagram represents a photograph of a soccer field. Construct on it the centerline and the two lines parallel to the centerline which divide the soccer field into three parts of equal area.

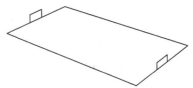

56. $ABCD$ and $KLMN$ are quadrilaterals. The points of intersections of AB and MN, BC and NK, CD and KL, DA and LM as well as AC and LN all lie on the same line. Prove that the point of intersection of BD and KM also lies on this line.

57. Prove that for any three convex bodies in space with no common point, there exists a plane such that their projections onto this plane are also have no common point.

58. Prove that $\sqrt{1 + \sqrt{2 + \sqrt{2^2 + \sqrt{2^3 + \cdots + \sqrt{2^n}}}}}$ is bounded above for any positive integer n.

59. A binary operation $\odot$ on the set $\{a, b, c, d, e, f, g, h\}$ is associative, has e as the identity and every element has an inverse. In other words, $(x \odot y) \odot z = x \odot (y \odot z)$ for any x, y and z, $e \odot x = x \odot e = x$ for any x, and for any x, there exists y such that $x \odot y = y \odot x = e$. Complete the operation table for $\odot$ shown in the diagram.

$\odot$	e	a	b	c	d	f	g	h
e								
a				e		g		d
b				a				
c								
d		h	g	f				
f				g				
g				h	e			
h						e		b

60. The diagram represents the map of a park showing its paths. A visitor starts walking from the center of the park, choosing one of the four paths at random. When she reaches another point where paths cross, she choose one of the four at random. She continues until she arrives at the edge of the park. What is the probability that she will be at a corner of the park?

61. Let $n > 4$ be an integer. Prove that at least $\binom{n-3}{2}$ convex quadrilaterals are determined by n points in the plane, with no three in a line.

62. Determine the number of different non-degenerate triangles with side lengths which are different positive integers not greater than n.

63. The equation $ax^4 + 4bx^3 + 6cx^2 + 4dx + e = 0$ has four real roots p, q, r and s. Prove that

$$(p - r)(q - s) + (p - s)(q - r) = 0$$

if and only if the determinant $\begin{vmatrix} a & b & c \\ b & c & d \\ c & d & e \end{vmatrix} = 0$.

64. We have a cardboard model of a cube and a cardboard model of a regular octahedron. They are oriented so that each vertex of the octahedron corresponds to a face of the cube. The edge joining two vertices of the octahedron then corresponds to the edge separating the corresponding faces of the cube. The model of the cube is cut along seven of its edges so that its six faces remain connected and can be laid out on a plane. Prove that if the model of the octahedron is cut along the five edges not corresponding to the cut edges of the cube, then its eight faces remain also remain connected and can be laid out on a plane.

65. (a) Prove that $\lim\limits_{t \to 0} \dfrac{x^t - 1}{t}$ is a real number.

    (b) Prove that $f(uv) = f(u) + f(v)$ for all real numbers u and v, where $f(x) = \lim\limits_{t \to 0} \dfrac{x^t - 1}{t}$.

66. The pairwise distances of infinitely many points on the plane are all rational. If they do not all lie on a line, it is possible that no three of them lie on a line?

67. Let $n \geq k$ be positive integers. A table-tennis tournament for $2n$ participants lasts $\binom{2n}{2}$ days. Each day, k pairs of participants are formed, playing against their partners in the pairs. No two days have the same k pairs, but this will not be true if the tournament is extended for another day. What are the possible values of n?

68. Let O be an arbitrary point inside a tetrahedron $ABCD$. Prove that

$$\angle AOB + \angle BOC + \angle COD + \angle DOA + \angle AOC + \angle BOD > 540°.$$

69. Determine all three-digit numbers each of which is equal to the sum of the cubes of its digits.

70. Let a_1, a_2, b_1 and b_2 be arbitrary numbers. For $n \geq 2$, let $a_{n+1} = 2a_n + a_{n-1}$ and $b_n = 3b_n - b_{n-1}$. Define $s_n = a_n + b_n$ for all n. Can s_n be expressed in terms of s_1, s_2, ..., s_{n-1}?

71. Among six points in the plane, no three collinear, prove that there exist three which determine a triangle having an angle not greater than $30°$.

72. D, E and F are points on the sides BC, CA and AB, respectively, of triangle ABC. If $AB + BD = AC + CD$, $BC + CE = BA + AE$ and $CA + AF + CB + BF$, prove that AD, BE and CF are concurrent.

73. (a) Let $\{a_n\}$ be a non-decreasing sequence of real numbers with $a_1 \geq 1$. Prove that for any positive integer

$$n, 0 \leq \sum_{k=1}^{n} \left(1 - \frac{a_{k-1}}{a_k}\right) \frac{1}{\sqrt{a_k}} < 2.$$

(b) Let c be any real number such that $0 \leq c < 2$. Prove that there exists a non-decreasing sequence of real numbers $(a_1 \geq 1)$ such that $\sum_{k=1}^{n} \left(1 - \frac{a_{k-1}}{a_k}\right) \frac{1}{\sqrt{a_k}} > c$ for infinitely many positive integer n.

74. In the plane are 100 points, no three collinear. Prove that among the triangles determined by three of these points, at most 70% of them are acute.

75. Does there exist a fifth degree polynomial f such that $f(-1) = 0$, $f(1) = 0$, for $-1 < x < 1$, the maximum value of $f(x)$ is 1, the minimum value of $f(x)$ is -1, and each of these two values is attained twice?

76. Given in the plane are two points A and B and a line. Construct the points P on this line such that the ratio $\frac{PA}{PB}$ are maximum and minimum respectively.

77. Peter has 12 white marbles and 8 black marbles in an urn. Paul has 8 white marbles and 12 black marbles in an urn. In each move, each draws at random a marble from his urn and puts it into the other urn. What is the probability that in the nth move, Peter draws a white marble and Paul draws a black marble?

78. Given n points in the plane, no three collinear, prove that they determine at least $\frac{1}{9}\binom{n}{2}$ and at most $\frac{1}{3}\binom{n}{2}$ triangles which do not have a common side.

79. In a country, every road is straight, every two roads intersect and no three roads intersect at the same point. At each point of intersection, one road passes under the other. Prove that it is possible to arrange that when traveling along any road, we pass over and under other roads alternately.

80. Given three circles in the plane, does there exist a circle of inversion such that the images of these circles are circles

    (a) with collinear centers;

    (b) with collinear centers and with two of the radii equal?

81. Prove that among 9 points in the plane, no three collinear, there exist 5 of them which determine a convex pentagon.

82. In a table-tennis tournament, each of five participants plays every other participants once, resulting in $\binom{5}{2} = 10$ matches. The order of these matches are determined at random. For $3 \leq n \leq 10$, what is the probability that after n matches, there exist three players who have played one another?

83. The circle $x^2 + y^2 = r^2$ intersects the curve $x^{\frac{2}{3}} + y^{\frac{2}{3}} = 1$ in eight points which determine two squares. Determine the value of r if each side of either square is tangent to the curve (not the circle) at one of the endpoints of the side?

84. Prove that there exist infinitely many rectangular blocks such that all edges and space diagonals have integral lengths, and of the three edges at each vertex, two have equal lengths while the length of the third differs from this common length by 1.

85. Prove that there does not exist a positive integer n such that the number of positive integers less than n and relatively prime to n is exactly $n - 10$.

86. There is a sufficient supply of fuel in a town, but each car there can carry just enough fuel to cover exactly half the distance to the capital. To get a car from the town to the capital, we use other cars which can transfer fuel to other cars along the way. However, they must have enough fuel to get back to the town. What is the minimum number of cars we must use, in addition to the one going to the capital?

87. Does there exist a set of eight points in the plane such that the perpendicular bisector of any two points in the set passes through exactly two other points in the set?

88. Construct a circle ω inside a given circle γ but not concentric with it, such that there exists a chain of eight circles in which each is tangent to both neighbors as well as to both ω and γ.

89. Find an arithmetic progression consisting of ten prime numbers.

90. How many permutations of $1, 2, \ldots, n$ are there such that each number differs from its position by at most 1?

91. D, E and F are the respective midpoints of the sides BC, CA and AB of triangle ABC. P, Q and R are points on EF, FD and DE respectively. If $DE + EP = DF + FP$, $EF + FQ = ED + DQ$ and $FD + DR = FE + ER$, prove that each of DP, EQ and FR passes through the incenter of ABC.

92. A_1C and BD are perpendicular diameters of a circle ω with center O. P is a point on the extension of OA_1. The tangent to ω at C intersects the line PD at Q and the line through P perpendicular to PB at R. The perpendicular bisector of OP intersects ω at A_2 and A_7. Prove that if $QR = OA_1$, then A_7, A_1 and A_2 are three consecutive vertices of a regular 7-gon inscribed in ω.

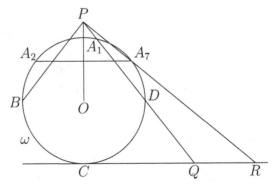

93. Is it true that for any positive integer n, there exist n consecutive positive integers each of which is divisible by the square of an integer greater than 1?

94. Evaluate $\lim\limits_{n\to\infty} \sum\limits_{k=0}^{\infty} \binom{n}{2k} 2^k \Big/ \sum\limits_{k=0}^{\infty} \binom{n}{2k+1} 2^k$.

95. The $4n^2$ points in an $2n \times 2n$ array are painted such that there are n red points and n blue points in each row and in each column. Prove that the number of line segments joining two adjacent red points in the same row or in the same column is equal to the number of line segments joining two adjacent blue points in the same row or in the same column.

96. The lengths of the sides of a cyclic quadrilateral are 36, 91, 315 and 260.

    (a) Compute its circumradius.

    (b) Prove that it has an incircle and compute its inradius.

97. Let $s_n = a_1 + a_2 + \cdots + a_n$ for some infinite sequence $\{a_n\}$ of positive numbers. Suppose that s_n tends to infinity as n tends to infinity. Prove that $\frac{a_1}{s_1} + \frac{a_2}{s_2} + \cdots + \frac{a_n}{s_n}$ also tends to infinity as n tends to infinity.

98. Let $a_1 = 2$ and $a_{n+1} = a_n^2 - a_n + 1$ for $n \geq 1$.

    (a) Prove that the terms of $\{a_n\}$ are pairwise relatively prime.

    (b) Prove that $\frac{1}{a_1} + \frac{1}{a_2} + \cdots + \frac{1}{a_n}$ tends to 1 as n tends to infinity.

99. Use the coordinate planes to divide the set of points (x, y, z) in space such that $|x| + |y| + |z| + |x + y + z| \leq 2$. Prove that among the resulting parts, six of them are congruent to one another, and any two of these six can be put together to form a cube.

100. A cyclic quadrilateral also has an incircle. Each side is arbitrarily a direction and converted into a vector. Prove that the sum of these four vectors is either the zero vector, or is perpendicular to the line joining the circumcenter and the incenter of the quadrilateral.

101. Inscribe an ellipse of maximum area into the area bounded by the parabolas $2py = x^2 - p^2$ and $2py = p^2 - x^2$.

102. Let f be a mapping which assigns a unique point $f(P)$ in the plane to every point P in the plane. Suppose that the distance between $f(P)$ and $f(Q)$ is equal to PQ whenever $PQ = 1$. Prove that the distance between $f(P)$ and $f(Q)$ is always equal to PQ.

103. Let Q_1 be a point on the x-axis and P_1 is the point of intersection of the vertical line through Q_1 and the parabola $y = x^2 + 1$. For $k \geq 1$, let Q_{k+1} be the point of intersection of the x-axis with the tangent to the parabola at P_k, and let P_{k+1} be the point of intersection of the vertical line through Q_{k+1} with the parabola. If $P_1 = P_7$, determine the number of different sets $\{P_1, P_2, \ldots, P_6\}$.

104. Let ABC be a triangle and let P be a point on its circumcircle. The line through P perpendicular to BC intersects BC at D and the circumcircle again at K. The line through P perpendicular to CA intersects CA at E and the circumcircle again at L. The line through P perpendicular to AB intersects AB at F and the circumcircle again at M.

It is known that D, E and F are collinear, and the line is called the Simson line of ABC with respect to P. Prove that the Simson lines of ABC with respect to K, L and M form a triangle similar to ABC but in the opposite orientation.

105. Let a be any six-digit number. Let $a = 10^k b_k + c_k$ for $k = 1, 2, 3, 4, 5$, where $c_k < 10^k$. Determine all positive integers n such that whenever it divides a, it also divides $10^{n-k} c_k + b_k$ for $k = 1, 2, 3, 4, 5$.

106. Let f be a positive function on the interval (0,1) such that for any real number λ in (0,1),

$$f(\lambda x_1 + (1 - \lambda)x_2) \geq \lambda f(x_1) + (1 - \lambda)f(x_2).$$

Moreover, f is integrable over (0,1) and the value of the integral is 1. What are the possible values of the integral of $f^2(x)$ over (0,1)?

107. Let $ABCD$ be a convex quadrilateral such that for any points P on AB and Q on CD such that $\frac{AP}{PB} = \frac{CQ}{QD}$, then the line PQ bisects the area of $ABCD$. Prove that $ABCD$ is a parallelogram.

108. In space are finitely many points, no three on a line. If a plane passing through any three of them contains at least one more of these points, prove that all these points lie on the same plane.

109. Let a and b be positive integers and let $\{x_n\}$ be a sequence of positive integers. What is the value of

$$\lim_{x \to \infty} \sum_{k=0}^{n} x_k a^k \Big/ \sum_{k=0}^{n} x_k b^k ?$$

110. A treasure is hidden inside one of n caves. For $1 \leq k \leq n$, the probability that the treasure is inside the kth cave is p_n and the time it takes to search the kth cave is t_k. Let $\{a_1, a_2, \ldots, a_n\}$ be a permutation of $\{1, 2, \ldots, n\}$. If we search the caves in the order determined by this permutation, $\sum_{k=1}^{n} p_{a_k}(t_{a_1} + t_{a_2} + \cdots + t_{a_k})$ is the total time required. Determine the permutation for which this sum is minimum.

111. A convex polyhedron P_1 has nine vertices A_1, A_2, $\ldots$, A_9. For $2 \leq k \leq 9$, the polyhedron P_k is obtained by translating P_1 so that the new position of A_1 coincides with the old position of A_k. Prove that at least two of these nine polyhedra have common interior points.

112. A line m is perpendicular to two lines ℓ_1 and ℓ_2 and intersect them at O_1 and O_2 respectively. Determine the locus of a point P in the plane such that a line may be drawn through P, intersecting ℓ_1 and ℓ_2 at Q_1 and Q_2 respectively, such that $O_1Q_1 \cdot O_2Q_2 = 1$ and Q_1 and Q_2 are

(a) on the same side of m;

(b) on opposite sides of m.

113. Let $a_n = 1$ and for $n \geq 1$, let $a_{n+1} = \sqrt{2\sqrt{3\sqrt[5]{5a_n}}}$. Is $\{a_n\}$ a convergent sequence?

114. Let f be a function on the interval $[0,1]$ such that for any real number λ in $(0,1)$,

$$f(\lambda x_1 + (1 - \lambda)x_2) \leq \lambda f(x_1) + (1 - \lambda)f(x_2).$$

Determine the set of ordered pairs (a, b) of real numbers such that the system of equations $f(\sin^2 x) + f(\cos^2 y) = a$ and $\sin^2 x + \cos^2 y = b$ has solutions.

115. Prove that $\frac{1}{2}(\tan x + \tan nx - \tan(n-1)x) > \frac{1}{n}\tan nx$ where n is a positive integer and x is a positive real number less than $\frac{\pi}{2n}$.

116. Let $a_1, a_2, \ldots, a_{2n}$ be real numbers with sum 0 such that for $1 \le i \le n$, $a_i = a_{n+i}$. Prove that there exists a positive integer k such that $a_k + a_{k+1} + \cdots + a_{k+j}$ is non-negative for $0 \le j \le n$.

117. O is a point inside triangle ABC. The lines OA, OB and OC intersect BC, CA and AB at K, L and M respectively. D, E, F, P, Q and R are the respective midpoints of BC, CA, AB, OA, OB and OC. Prove that D, E, F, K, L, M, P, Q and R lie on an ellipse.

118. S is a set of point in the plane such that Whenever A, B and D are points in S, then the point C such that $ABCD$ is a parallelogram also belongs to S. Suppose that for any positive number ϵ, there exists a circle in the plane with radius ϵ such that it contains infinitely many points of S. Prove that there exist an infinite number of points of S in any circle in the plane.

119. A number from the interval $[0,1]$ is chosen at random. What is the probability that the digit 5 appears among the first n decimal places of this number?

120. Prove that there exist infinitely many numbers of the form $2^n - 3$ which are pairwise relatively prime.

121. Prove that $\sum_{k=0}^{n}(n - 2k)^2\binom{n}{k} = n2^n$ for any positive integer n.

122. Let n be an integer greater than or equal to 5. From among the first $n - 1$ positive integers, choose more than $\frac{n+1}{2}$ of them. Prove that there exist two chosen numbers whose sum is also a chosen number.

123. The alphabet of a native tribe has only two letters, X and O. Every two words of the same length must differ in at least three places. Prove that the native language has at most $\frac{2^n}{n+1}$ words of length n.

124. Inside a 1×1 square is a polygonal line with at least 1000 segments of length 1 which do not intersect one another. Prove that there exists a line parallel to the sides of the square which intersects this polygonal line in at least 501 points.

125. In a table-tennis tournament, every participant plays a game against every other participant. Prove that there exists a participant A such that for any other participant B, either A beats B or A beats a third participant C who beats B.

126. The faces of three fair cubical dice are marked with the positive integers 1 to 18, using each number once. Is it possible that it is more likely to obtain a higher die-roll with the first die than with the second die, with the second die than with the third, and with the third than with the first?

127. Prove that for any positive integer n, there exists a finite set of points in the plane such that each point in the set is at a distance 1 from exactly n other points in the set.

128. Prove that if each face of a polyhedron has a center of symmetry, then the polyhedron itself has a center of symmetry.

129. Prove that $\binom{2n}{0}^2 - \binom{2n}{1}^2 + \binom{2n}{2}^2 - \cdots + \binom{2n}{2n}^2 = (-1)^n \binom{2n}{n}$ for any positive integer n.

130. Prove that there exists a multiple of 2^{100} in which every digit is 1 or 2.

131. Let a, b and c be real numbers such that $|a|+|b|+|c| \le \sqrt{2}$. Prove that $x^4 + ax^3 + bx^2 + cx + 1 = 0$ has no real roots.

132. A triangle can be constructed from any 3 of 5 given line segments. Prove that at least 1 of these 10 triangles is acute.

133. Let $f(x)$ be a polynomial of degree $n \ge 2$ such that all its roots are real. Let c be a real number such that $f'(c) \ne 0$. Prove that $f(x)$ has a root r such that $|r - c| \le n|\frac{f(c)}{f'(c)}|$.

134. There are 17 points in the plane, no three collinear. Each segment joining two of these points is painted in one of three colors. Prove that there exist three points such that the three segments joining them in pairs are all of the same color.

135. Prove that $(y+z)(z+x) + (z+x)(x+y) + (x+y)(y+z)$ is less than or equal to $PB \cdot PC + PC \cdot PA + PA \cdot PB$, where x, y and z are the distances from a point P inside triangle ABC to the sides BC, CA and AB respectively.

136. Prove that in every convex polygon, there exist three consecutive vertices such that the circle passing through them covers the polygon.

137. Find a differentiable function in the interval $(-1, 1)$ which has a maximum at 0 but its derivative does not change signs there.

138. String together the consecutive powers of 3, namely, 3, 9, 27, ... into a decimal expansion 0.3927 .... Prove that this number is irrational.

139. The points A, B, C_1, C_2, C_3, D_1, D_2, D_3, E_1, E_2 and E_3 are such that ABC_1D_1, ABC_2D_2, ABC_3D_3, $AC_1C_2E_3$, $AC_2C_3E_1$, $AC_3C_1E_2$, $AD_1D_2E_3$, $AD_2D_3E_1$ and $AD_3D_1E_2$ are all cyclic quadrilaterals. Prove that A, E_1, E_2 and E_3 are either collinear or concyclic.

140. Prove that a polygon can be dissected into parallelograms if and only if it has a center of symmetry.

141. Let $f(x)$ and $g(x)$ be functions defined on $[0,1]$. Prove that there exist numbers a and b in $[0,1]$ such that

$$|f(a) + g(b) - ab| \geq \frac{1}{4}.$$

142. Let n and k be positive integers and x be a real number in $[0,1]$. Prove that $(1 - (1 - x)^n)^k \geq 1 - (1 - x^k)^n$.

143. A circular arc of length 1 encloses with the chord joining its endpoints a circular segment. Determine the radius of the circle if the area of this circular segment is maximum.

144. A line ℓ through the center O of an equilateral triangle ABC and not passing through any vertex cuts the lines BC, CA and AB at D, E and F respectively.

    (a) Prove that the circle with center D and radius AD, the circle with center E and radius BE, and the circle with center F and radius CF, all pass through two points.

    (b) Find the locus of these two points if ℓ is rotated about O.

145. Find the maximum difference between two positive integers such that the digit-sum of each is a multiple of 13 but the digit-sum of any integer between them us not a multiple of 13.

146. Lamps A and B are both on initially. If we press button A, lamp A changes if necessary to the same status as lamp B. If we press button A again, lamp A changes if necessary to the opposite status as lamp B. The effect of pressing button A alternates thereafter. If we press button B and lamp A is off, then lamp B changes to the opposite status. If we press button B and lamp A is on, then lamp B remains in the same status. How can we turn off both lamps?

147. There are ten lattice points inside the triangle whose vertices are (0,0), (6,0) and (0,6). Each of them is joined to both (6,0) and (0,6). Find an approximate value for the average measure of the ten angles so formed.

148. Dissect a $36° - 36° - 108°$ triangle into the smallest number of acute triangles.

149. There are $2^n + 1$ positive real numbers on a blackboard. The product of each subset of $n + 1$ of these numbers is divided by the product of the remaining numbers, and the quotient is written on a whiteboard. Prove that the average of the numbers on the whiteboard cannot be less than the average of the numbers on the blackboard.

150. Each vertex of a regular n-gon is painted red or green. Each side or diagonal joining two vertices of the same color are painted yellow, while each side or diagonal joining two vertices of different colors are painted blue. The number of yellow sides is equal to the number of blue sides. The number of yellow diagonals of the same length is equal to the number of blue diagonals of that length.

(a) Prove that n must be the square of an integer.

(b) Does such a coloring scheme exist for $n = 16$?

151. Evaluate

$$\lim_{x \to \infty} \frac{\sqrt{n^2 - 1^2} + \sqrt{n^2 - 2^2} + \cdots + \sqrt{n^2 - (n-1)^2}}{n^2}.$$

152. Prove that if all dihedral angles of a tetrahedron are acute, then each face is an acute triangle.

153. Let r be a real number, $0 < r < 1$. Suppose $f_1(x) = 1$ if $0.5 \le x < 0.6$ and $f_1(x) = 0$ otherwise. For $n \ge 1$, $f_{n+1}(x) = f_n(x)$ if $f_n(x) > 0$. Suppose $f_n(x) = 0$. If the nth digit in the decimal expansion of x is 5, then $f_{n+1}(x) = r^{n-1}$. Otherwise, $f_{n+1}(x) = 0$. What is the value of $\int_0^1 f_n(x)dx$ as n tends to infinity?

154. One of the digits of the first n positive integers is chosen at random. What is the probability that this digit is 0 as n tends to infinity?

155. A turtle moves along a straight line for 6 minutes. At any moment during this period, the motion of the turtle is observed by at least one scientist. Each scientist observes the turtle for exactly 1 minute, during which the turtle moves forward 1 meters. Explain how it is possible for the turtle to have advanced 10 meters during these 6 minutes.

156. Join a pair of opposite edges of an 8×8 chessboard so that a cylindrical chessboard is formed. Prove that it is impossible to place counters on 8 of the squares so that there are no two in the same row, no two in the same column and no two in the same diagonal.

157. What are the maximum and minimum values of the difference between the probability of the occurrence of four arbitrary events simultaneously and the product of the probabilities of their individual occurrence.

158. Prove that for any positive integers n and k, there exists a unique set of positive integers $\{a_1, a_2, \ldots, a_k\}$ such that

$$n = \binom{a_1}{1} + \binom{a_2}{2} + \cdots + \binom{a_k}{k}.$$

159. Prove that the feet of at most two altitudes of a tetrahedron can lie outside the tetrahedron.

160. Prove that for any n points on the plane, there exists a point P on the plane such that for any line ℓ passing through P, there are at least $\frac{n}{3}$ of the points on either side of ℓ. Points on ℓ are considered to be on both sides of ℓ.

161. Determine all n-digit numbers such that each is equal to the nth power of its digit-sum.

162. In a chess tournament with $2n$ participants, each participant plays every other participants exactly once. Construct a schedule such that each participant plays exactly one game everyday.

163. When loot is shared between two pirates, one divides it into two shares and the other chooses. Each is satisfied that he has at least one half. Design a scheme when loot is shared among three pirates, so that each is satisfied that he has at least one third.

164. For any 4 of 100 points in space, there exists a sphere with radius 1 such that all 4 points are inside or on the surface of the sphere. Prove that there exists a sphere with radius 1 such that all 100 points are inside or on the sphere.

165. Let $a_0 = c$ for some c and for $n \geq 0$, let $a_{n+1} = 2(a_n - 1)^2$. Determine a_n in terms of c and n.

166. Prove that four pairwise disjoint spheres can be placed inside a large sphere with center O such that O is not inside or on the surface of any of them, but is not visible from any point on the surface of the large sphere.

167. No three of $2n$ points in the plane lie on a line. A line passing through two of them is said to be a bisecting line of these $2n$ points if it has exactly $n - 1$ points on each side of it. Prove that a line not passing through any of these points and has exactly $k \leq n$ points on one side of it will intersect exactly k bisecting lines.

168. Prove that the volume of a tetrahedron is at most $\frac{8}{9\sqrt{3}}$ of the cube of its circumradius.

169. Determine all real functions f which is bounded over some finite interval such that $f(1) = 1$ and $f(a+b) = f(a)+f(b)$ for all real numbers a and b.

170. Prove that every polynomial with real coefficients which does not take negative values can be expressed as the sum of the squares of two polynomials with real coefficients.

171. Prove that in an equilateral triangular lattice, there do not exist 4 lattice points determining a square.

172. Find all pairs (x, y) of real numbers such that

$$\cos x + \cos y - \cos(x + y) = \frac{3}{2}.$$

173. Prove that if all roots of the polynomial $f(x)$ are real, then all roots of $f(x) + \lambda f'(x)$ are also real for any real number λ.

174. Let $f(x)$ be a continuous function on $[0,1]$ with $f(0) = 0$ and $f(1) = 0$. Prove that for any integer $n > 1$, we have $f(x) = f(x + \frac{1}{n})$ where x is a real number x such that $0 \leq x < x + \frac{1}{n} \leq 1$.

175. In an equilateral triangular array, each number is 1, 2 or 3. Along one edge of the array, the number 1 does not appear. Along another edge, the number 2 does not appear. Along the third edge, the number 3 does not appear. Prove that there exists a unit equilateral triangle in the array consisting of three different numbers.

176. Solve $\sqrt{\lfloor\sqrt{\lfloor x\rfloor}\rfloor} = \left\lfloor\sqrt{\lfloor\sqrt{x}\rfloor}\right\rfloor$.

177. A chessplayer plays at least one game everyday and at most 13 games in a week. Prove that there exists a block of consecutive days in which the number of games played is exactly 1973.

178. Given in the plane are a circle and three points not on the circle. Construct a triangle with vertices on the circle and each side or its extension passes through one of the given points.

179. A solid is obtained by glueing six unit cubes, one on each face of a seventh unit cube, like three mutually perpendicular $1 \times 1 \times 3$ blocks intersecting one another at the central cube. Prove that space may be dissected into copies of this solid.

180. Prove that except for the powers of 10, every positive integer has a power whose first four digits are 1, 9, 7 and 4 in that order.

181. Consider the equation $15x + 6y + 10z = 1973$. How many integer solutions (x, y, z) are there with $x \geq 13$, $y \geq -4$ and $z \geq -6$?

182. P is a point inside a regular n-gon of side length 1 such that the distances from P to the sides are d_1, d_2, ..., d_n. Prove that $\frac{1}{d_1} + \frac{1}{d_2} + \cdots + \frac{1}{d_n} > 2\pi$.

183. On the surface of a sphere is a graph with edges intersecting only at vertices, and the degree of each vertex is 3. The regions are painted in red, yellow, blue and green such that regions sharing a common vertex have different colors. Prove that the total number of regions which have odd numbers of vertices and are painted in red or green is even.

184. Peter and Paul plays a game on a 3×3 table, taking turns putting a 0 or a 1 into a vacant square. Both 0 and 1 are available to both players. Peter wins if at any point, there are three 0s in a row or in a column. If this has not happened by the end of the game, Peter also wins if there are two identical rows or two identical columns. Otherwise Paul wins. Prove that no matter who goes first, Peter has a winning strategy.

185. Let $a_1 = 1$ and for $n \geq 1$, $a_{n+1} = 1 + \frac{n}{a_n}$. Determine the value of $\lim\limits_{n \to \infty} (a_n - \sqrt{n})$.

186. Let n be a positive integer. Peter and Paul plays a game with two piles of counters of sizes $\lfloor \frac{(1+\sqrt{5})n}{2} \rfloor$ and $\lfloor \frac{(3+\sqrt{5})n}{2} \rfloor$ respectively. Peter goes first, and turns alternate thereafter. At each turn, a player may take at least one counter from either pile, or the same number of counters from both piles. The player who takes the last counter wins. Prove that Paul has a winning strategy.

187. A convex polyhedron has at least as many vertices as faces. A pyramid is cut off at each vertex such that their bases have no common points. Prove that the resulting polyhedron has no insphere.

188. $\mathbf{OP}_1$, $\mathbf{OP}_2$, ..., $\mathbf{OP}_n$ are unit vectors in the plane such that P_1, P_2, ..., P_n all lie on the same side of some line passing through O. Prove that

$$|\mathbf{OP}_1 + \mathbf{OP}_2 + \cdots + \mathbf{OP}_n| \geq 1.$$

189. A soldier wants to verify that a field in the shape of an equilateral triangle is free of mines. He has a detector whose effect range is equal to half the altitude of the triangle. Starting from a vertex, what path should he follow if it has minimum length among all paths that would accomplish the task?

190. Let n be a positive integer and let a, b and c be the roots of $x^3 + nx + 7 = 0$. Prove that $a^7 + b^7 + c^7$ is the square of an integer.

191. Finitely many points in the plane are such that no three lie on a line, and any circle passing through three of them passes through a fourth point. Is it necessarily true that all points lie on a circle?

192. A hostess invites nine other guests to a party. She prepares an urn containing ten marbles each of which is red or white, and at least one of each color. Each of the ten people take turns drawing the marbles from the urn one at a time, the hostess going first. The first person who draws a red marble wins the door prize. How many white marbles should the hostess put in if she would like to minimize her own chance of winning the door prize?

193. Find the smallest positive integer such that $2^n - n$ is divisible by 61.

194. Does there exist a finite set S of points in space, not all on the same plane, such that for any two points A and B in S, there exist two points C and D in S such that AB and CD are distinct parallel lines?

195. Given in the plane are a circle and four points P, Q, R and S not on the circle. Construct a quadrilateral $ABCD$ with vertices on the circle such that the line AB passes through P, the line BC passes through Q, the line CD passes through R and the line DA passes through S.

196. Is it possible to tile the plane using only regular pentagons and regular decagons?

197. Let r be a real number in $(0,1)$ and let a_1, a_2, $\ldots$, a_n be positive numbers. Find positive numbers b_1, b_2, $\ldots$, b_n such that $a_k < b_b$ and $r < \frac{b_{k+1}}{b_k} < \frac{1}{r}$ for $1 \le k \le n$, while

$$b_1 + b_2 + \cdots + b_n < \frac{1+r}{1-r}(a_1 + a_2 + \cdots + a_n).$$

198. Let n be a positive integer and let B_1, B_2, $\ldots$, B_n be boxes with faces parallel to the coordinate planes. For any subset S of $\{1, 2, \ldots, n\}$, there exists a point P which belongs to B_k if and only if k belongs to S. What is the largest possible value of n?

199. A deck of 32 cards has eight cards of each of four suits. If 8 cards are drawn at random from this deck, what is the probability that there is at least card of each suit?

200. P is a point inside a tetrahedron $ABCD$. Prove that nine times the volume of $ABCD$ is less than or equal to

$$PA[BCD] + PB[CDA] + PC[DAB] + PD[ABC].$$

201. What is the locus of a point P such that four times the area of triangle ABC is equal to

$$PA^2 \sin 2A + PB^2 \sin 2B + PC^2 \sin 2C?$$

202. In rolling a standard six-sided die, what is the probability of rolling a 6 before rolling a 5?

203. An $a \times a$ square and a $b \times b$ square with no common points are inside a 1×1 square. Prove that $a + b \leq 1$.

204. A fair coin is tossed repeatedly. What is the probability of getting three heads in a row before getting two heads separated by exactly one tail?

205. Let k be any positive integer. Prove that we can always choose 2^k numbers from among the numbers 0, 1, ..., $3^k - 1$ such that no chosen number is the average of two other chosen numbers.

206. All entries in an $n \times n$ table are non-negative numbers. For each entry equal to 0, the sum of the other numbers in the same row plus the sum of the other numbers in the same column is at least n. Prove that the sum of all entries in the table is at least $\frac{n^2}{2}$.

207. Let e and f be disjoint ellipses. Prove that there exists a point E on e and a point F on f such that EF is perpendicular to the tangent to e at E, as well as to the tangent to f at F.

208. Three pirates are sharing loot. Pirate A divides the loot into three parts. Pirate B divides each part into three parts. Then pirate C takes one of the nine parts, followed by pirate A and then pirate B. This is repeated two more times. Is it possible for every pirate to feel that he has at least one third of the loot?

209. The alphabet of a native language has four letters, a, e, o and u. An effort is made to simplify the language. All appearances of e are deleted except in the single-letter word e. Whenever $aaaaaaa$, $ooooooo$ or $uuuuuuu$ appears, it will be replaced by e. Similarly, $aaau$, oou and $aaaao$ are replaced respectively by ua, uo and oa. Can the simplified language have 400 different words?

210. Prove that a square can be circumscribed about any bounded convex figure in the plane.

211. Let $ABCD - EFGH$ be a rectangular block. A plane through G intersects the edges AB, AD and AE or their extensions at P, Q and R respectively. How should the plane be chosen in order to maximize the volume of the tetrahedron $APQR$?

212. Find all polynomials which are the products of their derivatives with linear polynomials.

213. The sequences $\{a_n\}$ and $\{b_n\}$ are such that

$$\lim_{n\to\infty} \frac{a_n}{a_0 + a_1 + \cdots + a_n} = \lim_{n\to\infty} \frac{b_n}{b_0 + b_1 + \cdots + b_n} = 0.$$

Let $\{c_n\}$ be the sequence such that

$$c_n = a_0 b_n + a_1 b_{n-1} + \cdots + a_n b_0$$

for $n \geq 0$. Prove that $\lim\limits_{n\to\infty} \dfrac{c_n}{c_0 + c_1 + \cdots + c_n} = 0$.

214. Prove that if all vertices of a convex n-gon lie on the circumcircle of another convex n-gon, and the circumcenters of the two n-gons coincide.

215. Four points in space are not all on the same plane. How many different rectangular boxes have all of them as vertices?

216. What is the largest integer m such that the positive integers n, $n + 1$, $n + 2$, $\ldots, m$ may be divided into three groups such that the sum of any two different numbers in the same group is in another group?

217. A king visits every square of an 8×8 chessboard and returns to his starting square. Prove that the number of horizontal and vertical moves is at least 28.

218. No two sides of an n-gon are parallel. Prove that the number of lines through a given point and bisecting the area of the n-gon is at most n.

219. Given are an angle and a vector $\mathbf{v}$. A and B are points on a variable circle which is tangent to the arms of the angle such that $\mathbf{AB} = \mathbf{v}$. Determine the locus of the point B.

220. Let n be a positive integer. Prove that $\sum_{k=0}^{n} \binom{2n+1}{2k+1} 2^{3k}$ is not divisible by 5.

221. Let $a_1 = 1$ and for $n \geq 1$, $a_{n+1} = a_n + \frac{1}{a_1+a_2+\cdots+a_n}$. Prove that as n tends to infinity,

  (a) a_n tends to infinity;

  (b) $n(a_n^2 - a_{n-1}^2)$ tends to 2.

222. For what positive integer n does there exist in space an equilateral n-gon in which every two adjacent sides are perpendicular to each other?

223. In the plane are $4n$ points with no three on a line. Prove that it is possible to draw two perpendicular lines dividing the plane into four quadrants each of which contains exactly n of the points.

224. Let $a_1 = 1$ and for $n \geq 1$, $a_{n+1} = (2a_n)^{\frac{1}{a_n}}$. Does a_n approach a limit as n approaches infinity?

225. Each message consists of 7 digits each of which is 0 or 1. If during transmission, a single digit is reversed, then it is possible to determine uniquely the original message. What is the maximum number of different messages?

226. In a chess tournament, each participant plays one game against every other participant. At the end of the tournament, for any group of participants, one of them has played an odd number of drawn games against the others in the group. Prove that the total number of participants is even.

227. ABC is a triangle. A circle ω_1 is tangent to the lines AB and BC. The circle ω_2 is tangent to $\omega - 1$ and the lines BC and CA. The circle ω_3 is tangent to ω_2 and the lines CA and AB. The circle ω_4 is tangent to ω_3 and the lines AB and BC. Continue this way until ω_7 is constructed. Prove that if all these circles are inside ω_1, then ω_7 coincides with ω_1.

228. Let $P(x, y)$ be a polynomial of two variables. If there exists a real number $a > 1$ such that $P(x, a^x)$ is identically zero, prove that $P(x, y)$ is identically zero.

229. There are a white marbles and b black marbles with $a > b$. Marbles are drawn one at a time at random without replacement, until all are drawn. Prove that the probability of having drawn more white marbles than black marbles all the time is $\frac{a-b}{a+b}$.

230. Prove that a closed circular region cannot be divided into two congruent parts not having any points in common.

231. An infinite sequence of squares have respective side lengths a_1, a_2, a_3, ... such that the sequence $a_1^2 + a_2^2 + \cdots + a_n^2$ is convergent. Is it necessarily true that there exists a square such that all the squares in the sequence can be placed inside it without overlap?

232. Determine the largest possible value of the positive real number c such that the interval $(n\sqrt{2} - \frac{c}{n}, n\sqrt{2} + \frac{c}{n})$ does not contain any integer for any positive integer n.

233. Let a_0 be an arbitrary real number and $a_{n+1} = \sin a_n$ for $n \geq 0$. Determine in terms of a_0 the limit of $\sqrt{n}a_n$ as n tends to infinity.

234. The sequence $\{a_n\}$ consists of each of the positive integers exactly once in an arbitrary order. Prove that there exist positive integers $n > k$ such that a_n is between a_{n-k} and a_{n+k}.

235. What is the minimum volume of a cube whose vertices are all on the surface of a unit cube?

236. A 4×6 block of chocolate is to be cut into 24 individual pieces. What is the minimum number of cuts if only one piece of chocolate may be cut at a time?

237. At most how many determinants, each consisting of the same nine different numbers, can have different values?

238. Outside triangle $A_1B_1C_1$, equilateral triangles $B_1C_1D_1$, $C_1A_1E_1$ and $A_1B_1F_1$ are constructed. A_2, B_2 and C_2 are the respective midpoints of E_1F_1, F_1D_1 and D_1E_1. Triangle $A_3B_3C_3$ is obtained from $A_2B_2C_2$ in the same way, and so on. What can be said about $A_nB_nC_n$ as n goes to infinity?

239. $ABCD - EFGH$ is a unit cube. What is the minimum perimeter of a triangle with one vertex on each of the edges AE, BC and GH?

240. Let a_1, a_2, $\ldots$, a_n be non-negative real numbers. Prove that $\sum_{k=1}^{n} a_k^{n+1} \geq \sum_{k=1}^{n} a_k \prod_{k=1}^{n} a_k$.

241. The number of positive divisors of a positive integer n is denoted by $\tau(n)$. Prove that $\tau(1) + \tau(2) + \cdots + \tau(n)$ is equal to

$$2\left(\left\lfloor \frac{n}{1} \right\rfloor + \left\lfloor \frac{n}{2} \right\rfloor + \cdots + \left\lfloor \frac{n}{\lfloor \sqrt{n} \rfloor} \right\rfloor\right) + \lfloor \sqrt{n} \rfloor^2.$$

242. P is a point outside a circle with center O and $A_1 A_2 \ldots A_n$ is a regular n-gon inscribed in the circle. Prove that

$$\lim_{n \to \infty} \sqrt[n]{PA_1 \cdot PA_2 \cdots PA_n} = PO.$$

243. A_1, A_2, $\ldots$, A_n are fixed point inside a circle of radius r and P is a point moving along the circle. Prove that

   (a) $\max \sqrt[n]{PA_1 \cdot PA_2 \cdots PA_n} \geq r$.

   (b) $\min \sqrt[n]{PA_1 \cdot PA_2 \cdots PA_n} \leq r$.

244. For a fixed positive integer n, determine real numbers a_1, a_2, $\ldots$, a_n such that for all real number x,

$$x^n + a_1 x^{n-1} + \cdots + a_{n-1} x + a_n = (x - a_1)(x - a_2) \cdots (x - a_n).$$

245. Prove that $x^y + y^x > 1$ for all positive real numbers x and y.

246. Let $n \geq 5$ be an integer. Each number in an $n \times n$ table is 0 or 1. In each move, we may choose a 3×3 or a 4×4 subtable and interchange all 0s and 1s within the subtable. Is it always possible to get, in a finite number of moves, a table in which every number is 1?

247. Let the side lengths of a tetrahedron be a_1, a_2, $\ldots$, a_6 and the corresponding dihedral angles be θ_1, θ_2, $\ldots$, θ_6. Prove that $\dfrac{\pi}{3} < \dfrac{a_1\theta_1 + a_2\theta_2 + \cdots a_6\theta_6}{a_1 + a_2 + \cdots + a_6} < \dfrac{\pi}{2}$.

248. The digits of the primes numbers are written in their natural order after a decimal point. Prove that the resulting number 0.235711131719... is irrational.

249. What is the largest number of rays from a point in space such that every two of them determine an obtuse angle?

250. Let $a_0 = a_1 = 1$, $a_2 = 2$ and for $n \geq 2$, $a_{n+1} = a_n + a_{n-2}$.

    (a) Does $\frac{\ln a_n}{n+1}$ converge as n tends to infinity?

    (b) What can be said about the limiting value if the answer to (a) is yes?

251. A regular tetrahedron with side length 1 is projected onto a plane. Determine the maximum and minimum values of the area of the projection.

252. Initially, there is a group of two and a group of three rabbits. Each week, a new rabbit joins an existing group at random. If it joins a group of two, the group size increases to three. If it joins a group of three, the group splits into two groups of two. What is the probability that the nth rabbit joins a group of two?

253. The midpoints of the sides of a convex pentagon determines a smaller pentagon. Prove that its area is less than three-quarters and more than a half of the area of the larger pentagon.

254. Let $P(x)$ and $Q(x)$ be polynomials such that $P(x)$ is an integer if and only if $Q(x)$ is an integer. Prove that either $P(x) + Q(x)$ or $P(x) - Q(x)$ is constant.

255. Find four points in the plane such that their pairwise distances are at least 1 and the sum of the squares of their distances is minimum.

256. What is the smallest number of observations posts that we need on the surface of a sphere of radius 1 such that any unidentified flying objects at a distance 2 from the surface of the sphere should be observable from at least two observation posts?

257. Along the circle are placed 2^n digits each of which is 0 or 1. Starting from any digit, we obtain a binary sequence of length n by reading n consecutive digits in clockwise order. Is it possible for all 2^n sequences to be distinct?

258. What is the probability that among a group of 30 people, there are two having the same birthday?

259. Find a closed form for $\sin^4 \dfrac{2\pi}{n} + \sin^4 \dfrac{4\pi}{n} + \cdots + \sin^4 \dfrac{2n\pi}{n}$.

260. Is it possible to partition the positive integers into two sets neither of which contains an infinite geometric progression?

261. Evaluate $\lim\limits_{x \to \infty} x \left(\arctan \dfrac{x+1}{x+2} - \dfrac{\pi}{4} \right)$.

262. Let a and b be real numbers such that $0 < a < b < 1$. Let c_n be the number of positive integers n such that $a < n\sqrt{2} - \lfloor n\sqrt{2} \rfloor < b$. Prove that $\lim\limits_{n \to \infty} \dfrac{c_n}{n} = b - a$.

263. On a circle are 50 points distributed randomly. On another circle with equal radius are several arcs, also distributed randomly, such that their total length is less than $\frac{1}{50}$ of the circumference. Prove that the two circles may be superimposed on each other such that none of the 50 points is on any of the arcs.

264. A snail is at one end A of an elastic rope AB and the other end B is fixed. The snail crawls towards B at the speed of 1 centimeter per second while the rope lengths with A moving away from B at the speed of 10 meters per second. Can the snail ever reach B?

265. Let r be a real number. Let $a_1 = r$ and $a_{n+1} = r^{a_n}$ for $n \geq 1$. For what values of r does $\lim_{n \to \infty} a_n$ exist?

266. In a committee with five members, decision is made by the majority. One of the members makes mistakes 5% of the time. Three of them make mistakes 10% of the time. The last one makes mistakes 20% of the time.

   (a) What is the percentage of wrong decisions made by this committee?

   (b) What is the answer if the one who makes mistakes 20% of the time decides to vote along with the one with makes mistakes 5% of the time?

267. An $n \times n \times n$ block is composed of n^3 unit cubes, each of which is either black or white. For what value of n is it possible for each cube to share a common face with exactly two other cubes?

268. Is it possible to express the interval $[0,1]$ as a union of two sets A and B such that there exists a continuous function f on $[0,1]$ for which $f(a)$ is in B for all a in A, and $f(b)$ is in A for all b in B?

269. ABC is a triangle. D, E and F are the respective images of A, B and C across a line ℓ. P is a point on ℓ. The lines PD cuts the lines BC at P, the line PE cuts the line CA at Q and the line PF cuts the line AB at R. Prove that P, Q and R are collinear.

270. Construct a polygonal line with minimum length connecting the vertices of a given convex polygon.

271. Let $n > 1$ be an integer. Prove that there exists infinitely many positive integers which cannot be expressed as the sum of the nth power of n positive integers.

272. Prove that there does not exist a function f defined for all real numbers such that $|f(x) - f(y)| \geq \sqrt{|x - y|}$ for all real numbers x and y.

273. Prove that $\displaystyle\lim_{n \to \infty} \sqrt[n^2]{\binom{n}{0}\binom{n}{1}\cdots\binom{n}{n}} = \sqrt{e}$.

274. The distances between opposite edges of a tetrahedron are equal. Does it necessarily follow that the four altitudes of the tetrahedron are concurrent?

275. From the first 90 positive integers, five numbers are drawn at random. What is the probability that the difference between any two of them is at least 5?

276. The incircle of triangle ABC touches BC, CA and AB respectively. Prove that the orthocenter of DEF is collinear with the circumcenter and the incenter of ABC.

277. Let $\tau(n)$ denote the number of positive divisors of a positive integer n and $\sigma(n)$ their sum. Prove that

$$\sqrt{n} \leq \frac{\sigma(n)}{\tau(n)} \leq \frac{n+1}{2}.$$

278. R is the union of several circles. Prove that either one of the circles has area at least $\frac{1}{9}$ that of R, or two disjoint ones with total area at least $\frac{1}{9}$ that of R.

279. Let a_1, a_2, $\ldots$, a_n, b_1, b_2, $\ldots$, b_n, c_1, c_2, $\ldots$, c_n be non-negative real numbers. Let M denote the maximum of the sums $a_1+a_2+\cdots+a_n$, $b_1+b_2+\cdots+b_n$ and $c_1+c_2+\cdots+c_n$. Prove that $\displaystyle\sum_{i=1}^{n}\sum_{j=1}^{n}(a_ic_j + b_jc_i - a_ib_j) \geq M\sum_{k=1}^{n}c_k$.

280. Let m and n be positive integers. Prove that

$$\int_0^1 x^m(1 - x)^n dx = \frac{m!n!}{(m + n + 1)!}.$$

281. $ABCD - EFGH$ is a cube. A billiard ball is shot from A towards the midpoint of CF, and bounces off the inner walls of the cubes in the usual manner. Prove that it can never return to A.

282. A and B are two sets of integers such that B consists only of two numbers, and every integer can be expressed as the sum of a number from each of A and B. Prove that all non-zero integers which are not expressible as the difference of two numbers in A are odd multiples of the same integer.

283. In how many ways can k full bottles of milk, k half-full bottles of milk and k empty bottles be divided into three piles consisting of the same number of bottles and the same amount of milk?

284. The function f is differentiable on a closed interval $[a, b]$ and $f(a) = f(b)$. Prove that there exist two real numbers x and y in $[a, b]$ such that $f'(x) + 5f'(y) = 0$.

285. The function f is defined for all positive integers and takes only positive integral values. If $f(n + 1) > f(f(n))$ for all positive integer n, prove that $f(n) = n$ for all positive integer n.

286. A game called "Hex" is played on a playing board such that the one shown in the diagram, where $n = 5$ is the number of rows as well as diagonals from southwest to northeast. Anna goes first and places a red counter on a vacant space. Boris then places a green counter on a vacant space. Turns alternate thereafter. Anna wins as soon as there is a path of red counters linking the top edge to the bottom edge of the board, and Boris wins as soon as there is a path of green counters linking the left edge to the right edge of the board. Prove that this game can never result in a draw for any value of n.

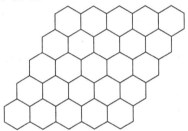

287. Consider the system of equations

$$a_{1,1}x_1 + a_{1,2}x_2 + \cdots + a_{1,n}x_n = 0,$$
$$a_{2,1}x_1 + a_{2,2}x_2 + \cdots + a_{2,n}x_n = 0,$$
$$\cdots = 0;$$
$$a_{n,1}x_1 + a_{n,2}x_2 + \cdots + a_{n,n}x_n = 0.$$

Solve the system if all coefficients are positive, the sum of all coefficients in each row and in each column is 1 and all coefficients in the main diagonal are equal to $\frac{1}{2}$.

288. There are 500 delegates in an international conference. They speak among them a total of $2n$ languages, with each speaking at least n of them. Prove that there exist 14 languages such that each delegate speaks at least one of them.

289. Let $a_1 \geq 2$ and for $n \geq 1$, $a_{n+1} = a_n^2 - 2$. Prove that

$$\frac{1}{a_1} + \frac{1}{a_1 a_2} + \frac{1}{a_1 a_2 a_3} + \cdots = \frac{a_1}{2} - \sqrt{\left(\frac{a_1}{2}\right)^2 - 1}.$$

290. A fair coin is tossed n times. Which is more likely, that no two heads appear consecutively, or no head appears in between two tails?

291. Find all polynomials P such that $P(x^2 - 2x) = (P(x-2))^2$.

292. The element in the ith row and jth column of a determinant is a^{ij} for some real number $a > 1$. Prove that the value of the determinant is not zero.

293. Starting with $S_0 = \{1, 1\}$, S_{n+1} is obtained from S_n by inserting between every two consecutive terms of S_n the sum of these two numbers. Thus we have $S_1 = \{1, 2, 1\}$, $S_2 = \{1, 3, 2, 3, 1\}$ and so on. How many times does the number 1978 appear in $S_{1000000}$?

294. A closed horizontal tube is 1 meter long. There are 100 marbles moving horizontally inside the tube, with the same speed of 10 meters per second. They will collide with one another and with the ends of the tube. How many collisions will happen in 10 seconds?

295. Prove that the number $5^{2^{n-1}} - 1$ has at least n prime divisors for every positive integer n.

296. In a table-tennis tournament, every participant plays every other participant once. At the end of the tournament, for any set of n participants, there exists a participant who has beaten them all. Prove that the total number of participants is at least $2^{n+1} - 1$.

297. Let a_1 be an arbitrary real number and let $a_{n+1} = (n+1)a_n$ for $n \geq 1$. Prove that if the infinite sequence $\{\sin a_n\}$ is convergent, then it has finitely many non-zero terms.

298. Construct a periodic function which does not have a smallest period and takes infinitely many different values.

299. The line joining the circumcenter and the centroid of a triangle bisects its area. Prove that the triangle is either an isosceles triangle or a right triangle.

300. Is it possible to place 19 counters on a 4×16 chessboard such that no two are on the same diagonal of length 2, 3 or 4?

301. Let $a_1 = 1$ and $a_{n+1} = a_n + \lfloor \sqrt{a_n} \rfloor$ for $n \geq 1$. Prove that a_n is the square of an integer if and only if $n = 2^k - k + 2$ for some integer k.

302. The following is an attempt to trisect an acute angle. Let the vertex of the angle be O and construct a semicircle with diameter AC along one arm of the angle, cutting that arm at A and the other arm at B. Let P and Q be points on the extension of AC such that $CP = OA$ and $CQ = \frac{2}{3}OA$. Let the lines BP and BQ intersect the semicircle at R and S respectively. Let T and U be points on CR and CS respectively such that $RT = SU = OA$. Let the line TU intersect AC at D. Then $\angle ADB$ is supposed to be one-third of $\angle AOB$. Estimate the accuracy of this construction.

303. Let a, b and c be real numbers not equal to 1. Prove that for any real number d, the equation $a^x + b^x = c^x + d$ has at most three real roots.

304. Let P be an arbitrary point on the circumcircle of a square $ABCD$.

    (a) Prove that the value of $PA^n + PB^n + PC^n + PD^n$ is constant for $n = 2$, 4 or 6.

    (b) For what other positive integer n is this true?

305. Let P be a point on a circle ω and ℓ be a line which does not intersect ω. For any two points A and B on the circle, we define a point C as follows and denote it by $A \odot B$. Let m be the line AB, or the tangent to ω at $A = B$. If ℓ and m intersect at some point Q, then C is the second point of intersection of ω with the line PQ. If m is parallel to ℓ, then C is the second point of intersection of ω with the line through P parallel to ℓ. Clearly, $A \odot B = B \odot A$ for any A and B. Prove that

    (a) $(A \odot B) \odot C = A \odot (B \odot C)$ for any A and B;

    (b) $P \odot A = A$ for any A;

    (c) for any A, there exists a B such that $A \odot B = P$.

306. On each square of a 99×99 chessboard is a beetle. Two beetles are said to be neighbors if the squares they occupy share at least one vertex. All the beetles leave the chessboard and then reoccupy it. Now two or more beetles may occupy the same square. If two beetles which are former neighbors either remain neighbors or move into the same square, prove that there is at least one beetle which reoccupies it original square.

307. There are two roads linking the towns A and B. Two trucks linked by a chain of length 20 meters travel from A to B on different roads, without breaking the chain. If one of them travel on a road from A to B and the other on the other road from B to A, each carrying a sphere of radius 11 meters, can they pass each other?

308. Let $\mathcal{F}$ be a family of subsets of an infinite set S such that any finite subset of S is the union of two disjoint subsets in $\mathcal{F}$. Prove that for any positive integer n, there exists a subset of S which is the union of two disjoint subsets of S in at least k different ways.

309. For any positive integer n, $\phi(n)$ denotes the number of positive integers less than or equal to n and relatively prime to n, while $\sigma(n)$ denotes the sum of all positive divisors of n. Determine all positive integer k for which $\phi(\sigma(2^k)) = 2^k$.

310. The lengths of each of the vectors $\mathbf{v}_1$, $\mathbf{v}_2$, ..., $\mathbf{v}_n$ is at most 1. Prove that we can choose the signs in the vector $\mathbf{v}_1 \pm \mathbf{v}_2 \pm \cdots \pm \mathbf{v}_n$ so that its length is at most $\sqrt{2}$.

311. Prove that $\lfloor \sqrt{p} \rfloor + \lfloor \sqrt{2p} \rfloor + \cdots + \lfloor \sqrt{kp} \rfloor = \frac{p^2-1}{12}$ if $p = 4k+1$ is prime.

312. Prove that $\displaystyle\int_0^\pi |\ln \sin x| dx \le \frac{\pi^3}{12}$.

313. Is it possible to arrange the positive rational numbers in a sequence and take an open interval of length $\frac{1}{n}$ around the nth number so that the union of these open intervals cover all positive real numbers?

314. Let m and n be relatively prime positive integers. Let $\{a_k\}$ be a sequence. Prove that if the sequences $\{a_{mk}\}$ and $\{a_{nk}\}$ are both arithmetic series, then so is $\{a_k\}$.

315. (a) Prove that for any integer $k > 2$, there are infinitely many positive integers n such that $k^n - 1$ is divisible by n.

    (b) What happens if $k = 2$?

316. A cubic polynomial with integral coefficients has three distinct real roots. Does that always exist a real number c such that $P(x) + c$ has three rational roots?

317. Solve the system of equations $x^3 + y^2 + z = 3$, $x + y^3 + z^2 = 3$ and $x^2 + y + z^3 = 3$.

318. Let $ABCD$ be a convex quadrilateral. Let A_1, A_2, $\ldots$, A_{n-1} be points on AB such that

$$AA_1 = A_1A_2 = \cdots = A_{n-1}B.$$

Let B_1, B_2, $\ldots$, B_{n-1} on BC, C_1, C_2, $\ldots$, C_{n-1} on CD and D_1, D_2, $\ldots$, D_{n-1} on DA be similarly defined. Joining corresponding points on opposite sides divides $ABCD$ into n^2 smaller quadrilaterals in an $n \times n$ configuration. Prove that the total area of the n quadrilaterals along a diagonal is equal to $\frac{1}{n}$ that of $ABCD$.

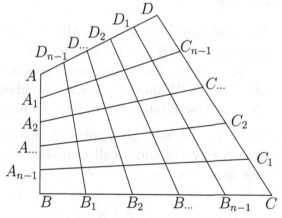

319. To each subset A of a set S, we associate a subset $f(A)$ of S such that $f(A) \subset f(B)$ if $A \subset B$. Prove that there exists a subset C of S such that $f(C) = C$.

320. A rectangle is partitioned into rectangles. If each of these rectangles has one side of integral length, then the original rectangle also has one side of integral length.

321. For which positive integer n does there exist a continuous function defined for all real numbers and taking on all values exactly n times?

322. Let $k > m \geq n$ be positive integers. Among any m of k given arcs on a circle, there exist n of them with a common point.

    (a) Prove that there exist $m - n + 2$ points such that each of the k given arcs contain at least one of these points.

    (b) Show an example where no such $m - n + 1$ points exist.

323. Does there exist a strictly increasing function defined for all real numbers such that its range consists only of irrational numbers?

324. (a) In a building with n floors, the elevator may be on any particular floor at any moment. A call for the elevator may come with equal probability from any of the floors. What is the average number of floors the elevator has to pass over before reaching the caller?

    (b) What is this average number if the building has a second elevator and a call is answered by the nearer elevator?

325. There are n red cards and n black cards in a thoroughly shuffled deck. The cards are turned over one at a time. Before turning over each card, a prediction of its color is made. What is the expected number of correct predictions?

326. ABC is a triangle with $AB \neq AC$. AD, BE and CF are the angle bisectors. If $DE = DF$, what are the bounds for the measure of $\angle CAB$?

327. Prove that it is impossible to dissect a rectangle into six squares of different sizes.

328. The first sequence consists of the positive integers from 3 on in their natural order. Each subsequent sequence is obtained from the preceding one by moving its first term k to the kth place. Prove that every integer greater than 3 appears as the first term of some sequence.

329. For any three points A, B and C inside a non-convex polygon, there exists a point P such that the segments PA, PB and PC all lie entirely inside the polygon. Prove that there exists a point Q such that for any point D inside the polygon, the segment QD lies entirely inside the polygon.

330. Each face of fair die in the shape of a regular icosahedron is labeled with an arbitrary positive integer. The die is rolled and the label of the top face is recorded. This is done three times. Prove that the probability that the sum of the three recorded numbers is divisible by 3 is at least $\frac{1}{4}$.

331. Prove that a rectangle can be dissected into squares of different sizes if and only if the ratio of its sides is rational.

332. There is a counter on each of 32 squares on an 8×8 chessboard, with 4 in each row and 4 in each column. Prove that it is possible to remove 24 of the counters so that 1 counter is left in each row and in each column.

333. Prove that for all integers $n \geq 1$, one can find a rational number smaller than 1 which cannot be expressed as a sum of the reciprocals of n positive integers.

334. For an arbitrary real number in the interval $[-1, 1]$, denote by $M(x)$ the maximum value at x of those quadratic polynomials which have non-negative values in the interval $[-1, 1]$ such that the area under the curves in this interval are equal to 1. Prove that $\min_{-1 \leq x \leq 1} M(x) = \dfrac{2}{3}$.

335. Let X and Y be subsets of the set of real numbers. Denote by $X \cdot Y$ the set which consists of those real numbers which are products of one real number from X and one real number from Y. Similarly, the elements of $X + Y$ are obtained by adding elements of X and Y. To which real numbers c does there exist a non-empty subset X of real numbers such that $X \cdot X + \{c\} = X$?

336. We have n oxygen cylinders of equal volume. Denote the pressures in them respectively by $p_1 \leq p_2 \leq \cdots \leq p_n$. In each move, we can connect the cylinders in an arbitrary group. The common pressure will be the average of the original pressures. The cylinders can then be disconnected. Find the maximum pressure which can be obtained in the first cylinder after a finite number of moves.

337. In a table tennis tournament, each of 2^n participants plays every other participants exactly once. Prove that there exist $n + 1$ players X_1, X_2, ..., X_{n+1} such that X_i beats X_j if and only if $i < j$.

338. Any two towns of a country are connected by direct bus or plane services. In whatever way we plan our journey, passing through $2k$ towns where $k \geq 3$, we must use both buses and planes. Prove that the same is true for a journey passing $2k + 1$ towns.

339. Prove that $2n < 4^{\frac{1}{n}} + 4^{\frac{2}{n}} + \cdots + 4^{\frac{n}{n}} \leq 3n$ for all integers $n > 1$.

340. Construct a one-to-one correspondence between the points and lines of the plane such that each point lies on the line corresponding to it.

341. There are 20 towns in a country. From each town, 4 other towns can be reached by direct flights. Prove that the flights can be distributed to two companies so that each company provides 2 direct flights from each town.

342. Does there exist a non-empty subset M of the real numbers such that for an arbitrary positive real number r and any $a \in M$, there is exactly one $b \in M$ such that $|a - b| = r$?

343. Does there exist a positive integer n such that $1, 2, \ldots, n$ can be divided into at least two arithmetic progressions with pairwise different common differences, each consisting of at least three terms and having no common elements?

344. The sequence $\{a_n\}$ is defined by $a_0 = a_1 = 1$ and for $n \geq 1$, $a_{n+1} = 2a_n + a_{n-1}$. Prove that $2(a_{2n}^2 - 1)$ is the square of an integer for every positive integers n.

345. Let $a_1, a_2, \ldots, a_n$ be distinct integers. For $1 \leq i \leq n$, let $b_i = (a_i - a_1) \cdots (a_i - a_{i-1})(a_i - a_{i+1}) \cdots (a_i - a_n)$. Prove that for all positive integer k, $\frac{a_1^k}{b_1} + \frac{a_2^k}{b_2} + \cdots + \frac{a_n^k}{b_n}$ is an integer.

346. Six counters of different colors are placed on the six dots in the diagram below. In each move, we may rotate one of the two circles 90° about its center, carrying along the four counters on it. Is it possible, after a finite number of moves, to obtain a configuration in which the three upper counters remain in place while the lower counter initially on the left has traded places with the lower counter initially in the center?

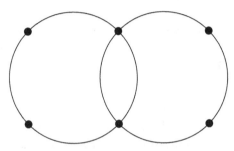

347. Five numbers are drawn at random from the first 90 positive integers. Determine the probability that each number is the square of an integer or the sum of the squares of two integers.

348. We have a right circular cone with base radius n and height n, and n rings each of height 1 and inner radius from 1 to n respectively. We have to place the rings on the cone one at a time, according to the order in which they are arranged. Each ring drops as far as possible down the side of the cone, stopped either by the cone itself or by other rings already placed. Thus if the rings are arranged in descending order of size, all of them can be placed, but if they are arranged in ascending order, only the smallest one can be placed. How many different final configuration can we obtain?

349. Given are 101 points in the plane, no three collinear. Paint each line segments connecting two of these points in one of n colors such that the three segments joining pairwise three of the points are either all of the same color or all of different colors. Prove that if $n > 1$, then $n \geq 11$.

350. A transformation of the plane carries lines into lines and circles into circles. Prove that a triangle whose sides are in the ratio 3:4:5 is transformed into a triangle similar to itself.

·351. As n tends to infinity, what is the limiting value of

$$\frac{1}{n^2}\left(\sin\frac{\pi}{n} + 2\sin\frac{2\pi}{n} + 3\sin\frac{3\pi}{n} + \cdots + (n-1)\sin\frac{(n-1)\pi}{n}\right)?$$

352. Suppose that the human race does not become extinct but each human being's life span is finite. Moreover, nobody will have infinitely many children. Prove that there will be an infinite sequence of human beings such that the first one is living now and each one after is a child of the preceding one.

353. Prove that $\lfloor nx \rfloor \geq \lfloor x \rfloor + \frac{\lfloor 2x \rfloor}{2} + \frac{\lfloor 3x \rfloor}{3} + \cdots + \frac{\lfloor nx \rfloor}{n}$ for all positive integers n and all real numbers x.

354. A building has ten floors numbered 0 to 9 from the bottom. The distance between floors is 4 meters. An elevator starts on floor 0 and stops on each floor once and only once. The stops must be at odd-numbered floors and at even-numbered floors alternately. What is the maximum distance the elevator has moved?

355. In (6,7,8) and (48,49,50), the prime divisors of 6 are the same as those of 48, the prime divisors of 7 are the same as those of 49, but the prime divisors of 8 are not the same as those of 50. Find two blocks of three consecutive integers for which we have the same prime divisors in all three cases.

356. In the sequences $\{a_n\}$ and $\{b_n\}$, $a_1 = 3$ and $b_1 = 1$. For $n \geq 3$, $a_n = \frac{a_{n-1}+a_{n-2}}{2}$ and $b_n = \frac{2b_{n-1}b_{n-2}}{b_{n-1}+b_{n-2}}$. Prove that $\lim\limits_{n\to\infty} a_n = \lim\limits_{n\to\infty} b_n$.

357. At most how many points can be placed inside a 1×1 square so that the distance between any two of them is greater than $\frac{1}{2}$.

358. A building has ten floors numbered 0 to 9 from the bottom. The distance between floors is 4 meters. An elevator starts on floor 0 and stops on each floor once and only once. The stops are chosen at random. What is the expected value of the distance the elevator has moved?

359. Determine all finite subsets S of the positive integers such that for any two numbers in S, not necessarily different, their sum or their absolute difference is also in S.

360. On a 5×5 chessboard, 19 squares are chosen. Each is then painted in one of at least 10 different colors, such that there are at most 2 squares of any color. We then place 5 rooks on the painted squares such that there are always two which are on squares in the same row, in the same column or of the same color.

   (a) Prove that the 19 squares can be chosen so that at least one square from each row and from each column is chosen.

   (b) Can the 19 squares be chosen in several essentially different ways?

361. In the plane are 20 painted points, no three on a line. There are 5 points in each of 4 colors. Prove that we can always find four non-intersecting segments joining pairs of points of the same color, with points of different colors joined by different segments.

362. A convex figure is inside a convex polygon in the plane. Prove that the perimeter of the figure is less than the perimeter of the polygon.

363. Let $\sum_{n=1}^{\infty} a_n$ and $\sum_{n=1}^{\infty} b_n$ be divergent series of positive numbers. Does it necessarily follow that $\sum_{n=1}^{\infty} \dfrac{2a_n b_n}{a_n + b_n}$ is also divergent?

364. Let C be a point on a smooth curve AB with a tangent at A. The following is a flawed proof that as C moves along the curve towards A, the ratio of the lengths of the segment AC and the curve AC tends to 1. "Let T be a point on the tangent to the curve at A, and let D be the point of intersection of BT with the extension of AC and let the curve AD be obtained from the curve AC by a homothety centered at A. Then this curve is also tangent to AT. Since the curve AD is inside triangle DAT, it is shorter than $AT + DT$. By the Triangle Inequality, $AT < AD + DT$. Hence the length of the arc AD is less than $AD + 2DT$. On the other hand, it is greater than AD. As C moves towards A along the curve AB, D moves along the segment BT towards T. The desired results follows since the length DT tends to 0." What additional assumption is made in this argument?

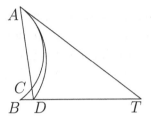

365. Does there exist a function f from the real numbers to the real numbers such that $\lim\limits_{x\to\infty} f(x) = \infty$ and

$$\lim_{x\to\infty} \frac{f(x)}{\ln(\ln\cdots(\ln x)\cdots)} = 0$$

for any number of natural logarithms in the denominator?

366. Determine all possible values of the radius of a circle such that their interior points can be painted in two colors, with the endpoints of any segment of length 1 painted in different colors.

367. Let p be a prime number. We compute the product of the elements of every subset of $\{1, 2, \ldots, p-1\}$. The product for the empty subset is taken to be 1. Prove that the sum of all these products is divisible by p.

368. Around the coast of Pagan Island are 26 villages. Their names start with the letters A, B, ..., Z in cyclic order. At various time in its history, it has been visited by 26 missionaries. Their names also start with the letters A, B, ..., Z. Each arrives at the village whose name starts with the same letter as his name. If the village is Pagan, it is converted the missionary moves clockwise along the coast to the next village. If the village has been converted already, the missionary is eaten, and the village becomes Pagan again. While there may be several missionaries on Pagan Island at the same time, no two appear (or disappear) together in the same village.

Clearly, all 26 missionaries leave their bones on Pagan Island. At most how many villages may remain converted by their combined effort?

369. Prove that there does not exist a polynomial p such that $p(k) = 2^k$ for infinitely many integers k.

370. Prove that in a string of letters of sufficient length, the probability that it contains the substring "WIZARD" is sufficiently close to 1.

371. At a party, each member knows at least k others. Prove that at least k of them can be seated at a round table so that each knows both neighbors.

372. A regular hexagon is cut up into polygons in such a way that at least 3 polygons meet in all corners of every polygon which are not on the sides of the hexagon. Prove that the average of the number of the sides of the polygons is not greater than 6.

373. Does there exist a multiple of 5^{100} which does not contain the digit 0?

374. In a party of 25 members, whenever two members do not know each other, they have common acquaintances among the others. Nobody knows all the others.

   (a) Prove that by adding the numbers of acquaintances of all members, the sum is at least 72.

   (b) Can the sum be 92?

375. (a) Inscribe a quadrilateral inside a quadrilateral of area 1 such that the areas of the triangles at the vertices are equal.

   (b) Prove that the area of the inscribed quadrilateral is at least $\frac{1}{2}$.

376. Do there exist two pentagons in the place such that each vertex of either pentagon lies on a line containing a side of the other pentagon?

377. A and B are the endpoints of a polygonal line and $AB = 1$. Prove that for all positive integer n, there exist two points on the polygonal line such that the segment joining them is parallel to AB and has length $\frac{1}{n}$.

378. (a) Do there exist two tetrahedra in space such that each vertex of either tetrahedron lies on a plane containing a face of the other tetrahedron?

   (b) Do there exist two such terahedra such that one if obtained from the other by a rotation?

379. C is the center of gravity of a bounded plane figure F with area t. A circle ω contains the circle with diameter AB for any two points A and B is R. Does ω necessarily contain the circle with center C and area $2t$?

380. A counter is placed on each square on an infinite chessboard below a certain horizontal line. In each move, a counter may jump over another counter on a square sharing a common side and lands on the square immediately beyond which must be vacant prior to the jump. The counter jumped over is removed. Prove that it is impossible to send a counter to the fifth row above the horizontal line after a finite number of moves.

381. Prove that among any n points inside an equilateral triangle of side length 1, there exist at most $\frac{n^2}{3}$ pairs whose distance is not greater than $\frac{3}{4}$.

382. Does there exist a monotonically decreasing function f with derivatives of all orders such that $\lim\limits_{x \to \infty} f(x) = 0$ but $\lim\limits_{x \to \infty} f'(x) \neq 0$?

383. The midpoints of all segments joining pairs of n points in the plane are painted red. How many red points can there be?

384. Peter and Paul take turns placing respectively black and white counters on an infinite chessboard, one counter per square. Peer goes first and turns alternate thereafter. The game is won by the player who succeeds first in having four counters of his color at the corners of a square and a fifth counter at the center of that square. There is no restriction on the size or orientation of this square. Prove that Peter has a winning strategy.

385. A hole is cut through a cube of edge length a so that a cube of edge length b can pass through the hole. What is the largest possible value for b?

386. Peter and Paul toss for a fair coin. If it lands heads, Peter gets one cookie from Paul. If it lands tails, Paul gets one cookie from Peter. Peter wins the first toss. What is the probability that during the first 100 tosses, Peter always has at least as many cookies as he has initially?

387. Let a and b be positive integers. Prove that $\frac{a}{b}$ can be expressed as a sum of at most a reciprocals of different positive integers.

388. Let $\{a_n\}$ be a sequence of positive integers such that for all positive integer n, $a_n < a_{n+1} + a_n^2$. Does it necessarily follow that the series $\sum_{n=1}^{\infty} a_n$ diverges to infinity?

For more information about KÖMAL, check out the websites:
https://www.komal.hu/home.e.shtml.
https://www.komal.hu/matfund/matfund.e.shtml.

Printed in the United States
by Baker & Taylor Publisher Services